THERMOSTATIC CONTROL
Principles and Practice

THERMOSTATIC CONTROL

Principles and Practice

by

V. C. Miles, C.Eng., F.I.Gas.E., M.I.M.C.

London
NEWNES – BUTTERWORTHS

THE BUTTERWORTH GROUP

ENGLAND
Butterworth & Co (Publishers) Ltd
London: 88 Kingsway, WC2B 6AB

AUSTRALIA
Butterworths Pty Ltd
Sydney: 586 Pacific Highway, NSW 2067
Melbourne: 343 Little Collins Street, 3000
Brisbane: 240 Queen Street, 4000

CANADA
Butterworth & Co (Canada) Ltd
Toronto: 14 Curity Avenue, 374

NEW ZEALAND
Butterworths of New Zealand Ltd
Wellington: 26–28 Waring Taylor Street, 1

SOUTH AFRICA
Butterworth & Co (South Africa) (Pty) Ltd
Durban: 152–154 Gale Street

First published in 1965 by George Newnes Ltd. Second edition published in 1975
by Newnes–Butterworths, an imprint of the Butterworth Group

ISBN 0 408 00131 3

Printed in England by The Whitefriars Press Ltd., London and Tonbridge

CONTENTS

PREFACE

OVER the last 40 years the measurement and control of temperature has become an increasingly important subject. In the domestic sphere, better living standards have led to the introduction of a number of new heating techniques, and the use of increasingly complex domestic appliances—for cooking, washing, heating and refrigeration—all of which have temperature control problems peculiar to themselves. In industry, the use of more complex plant and the advantages gained from the elimination of manual supervision have given rise to further demands for thermostatic control equipment.

A sizeable industry has arisen to provide the numerous thermostatic devices needed today, with the result that there is now available a vast range of thermostatic controls. This wide range can cause considerable confusion to those who have to deal with heating and cooling problems.

Whilst much has been written on the theory of automatic control, very little practical information on the subject of temperature-controlling devices has been published. The present book has been written to meet this need. It deals with the practical problems involved in the design and application of thermostatic control equipment so that the engineer, student or designer can see what the essentials of good practice are and be able to select and apply the right combination of controlling devices for a given problem.

As it is not assumed that the reader is conversant with the theory of control, the early chapters of the book are devoted to the basic principles of thermal elements and their operation. The later chapters deal with the practice of thermostatic control, concentrating on actual examples rather than theoretical analysis.

Many of the illustrations show particular manufacturers' products, but it must be pointed out that in most instances any one of several available would equally have served the purpose.

The author sincerely thanks his many friends in the industry who have helped in the preparation of the book by lending illustrations and offering much helpful advice, and records his indebtedness to Mr. E. Ower, on whose suggestion the book was first undertaken and who edited and revised the original manuscript. His patient attention to detail, suggestions and advice have saved the writer from many pitfalls and brought the book up to a standard that would not otherwise have been possible.

V. C. M.

PREFACE TO THE SECOND EDITION

SOME ten years have elapsed since the first edition of this book became available, since when there has been a considerable amount of progress in the instruments available in most sectors of the thermostatic control industry. The advent of commercially available electronics, improved techniques in the application of these devices and other improvements that have become available, have all contributed to the changes that have taken place.

In the last few years the whole industry has prepared itself for the metric era, and whereas five years ago the use of Fahrenheit temperature scales and Imperial measurement was universal, all new thinking must be directed towards the use of the Celsius or Centigrade scale and S.I. units.

All these aspects have been borne in mind in writing the second edition, and it is hoped that in its new form this book will continue to be of benefit to all those interested in this sphere of automatic control.

V. C. M.

INTRODUCTION

DESPITE the large number of thermostatic devices now available and the wide variety of their applications the control functions they are called upon to provide are common to all of them, and may be summarized thus:

(1) To control the temperature of a condition within certain limits of accuracy.
(2) To control a piece of equipment in such a manner that it is operated safely and economically.
(3) To provide continuous monitoring of a temperature condition that would be difficult or impossible by manual operation.
(4) To provide a safety margin as does a pressure safety valve, so that on an excessive rise in temperature, the sequence of operations providing the source of heat is brought either to a complete standstill until manually reset, or to a temporary halt until safe temperature conditions again prevail.

Thermostatic control now covers such a wide variety of devices that it is becoming increasingly difficult to lay down a general definition covering them all. It is, however, possible to state the three basic essentials of a complete thermostatic device, namely: (1) the prime mover; (2) the ranging and transmission component; and (3) the control mechanism.

PRIME MOVER

The *prime mover* consists of a material or combination of materials, solid, liquid, or gaseous, which change their volume or physical characteristics in relation to temperature, and can repeat these changes consistently many times over a defined range of temperature. This component is also known as the *sensing* or *thermal element*. Simple examples are lengths of metal or capsules enclosing liquids or gases, in which temperature changes produce changes in length, volume, or pressure respectively. These and many other elements can be used successfully in thermostatic controls.

The selection of a prime mover for any particular thermostatic control might at first sight seem to be comparatively simple, since many forms of this device are often commercially obtainable for a given temperature range. But when all the various factors that govern the choice of a prime mover have been assessed only one or two of the choices available will usually be suitable for any particular application.

Some of the more important factors to be considered are power output, effective temperature range, and power per temperature change.

Power output

Each thermal prime mover has a certain amount of work to do, e.g. either to operate a very small switch mechanism, or to operate a comparatively large valve which in turn has to overcome an opposing pressure of fluid or gas. Therefore, it must first be decided whether an element is available which will do the work required of it, and, if not, what servo or amplifier mechanism may be required to supplement it.

Effective temperature range

A decision must next be made regarding the total range of temperature over which thermostatic control is required. Under this heading must also be considered whether any over-run or under-run conditions are likely to apply. A thermal element may be perfectly satisfactory for operation within the temperature range for which the thermostat is designed, but may suffer permanent damage if subjected to temperatures above the selected range.

Power per temperature change

Apart from the temperature range required, it must be decided what work, or output, is required at any set point within the range. Some prime movers are capable of working over wide ranges of operating temperatures, but are incapable of providing a large amount of work at any particular set-point.* This sometimes precludes their use in direct-acting systems where the thermal element actually moves the controller unless servo or amplifier systems are introduced. On the other hand, other prime movers exist which can produce a considerable amount of work output for a small change in temperature, but are unsuitable for operating over a wide range of temperature points.

RANGING AND TRANSMISSION COMPONENT

This is the mechanical arrangement which presets the thermal device or the linkage attached to it so that, at a selected temperature, move-

* Definitions of this and a number of other common technical terms used in thermostatic control engineering are given in Chapter 6.

ment obtained from the thermal element can be transmitted to a control component which will then be regulated in relation to any further changes in temperature at the sensing element.

The temperature-selection mechanism is generally operated by a knob on the front of the thermostat marked in terms of degrees of temperature, so that the set-point or operating point of the thermostat can be easily preselected. The transmission mechanism becomes more involved if it has not only to transmit movement from the thermal element to the control mechanism, but also has to multiply it in order to obtain a useful movement over a small difference in temperature. The transmission mechanism may also incorporate devices for converting the slow movement of the thermal element into a positive snap action, as would be required when the thermal element has to control a substantial electric current (where a slow movement could lead to arcing and contact deterioration).

CONTROL MECHANISM

This is the final component in the thermostat. It receives the signal from the thermal element, via the ranging and transmission component, and reacts on receiving this signal to regulate the correcting medium. By correcting medium is meant any heating or cooling effect that is used to correct any departure from the desired value of the temperature being sensed by the thermal element. It may be gas, water, oil, or air, exerting a cooling or heating influence as demanded by the specific system to which the thermostat is applied, according to whether the temperature is higher or lower than that called for by the thermostat setting. Similarly, the control mechanism could take the form of a switch in an electric circuit which, in turn, controls valves or heaters larger than those which the power of the thermal element could control directly.

Finally, it could be an amplifier circuit in an electronic system, receiving and amplifying the signals from a resistance-type temperature-sensing element and passing on a more powerful signal capable of controlling heavy electrical loads or automatic valve gear.

These three essentials are to be found in any thermostat design, their complexity depending on the nature of the problem in hand. In simple devices, where space or economy are of greater importance than extreme accuracy, they can be combined to a point where there is no clear mechanical division between them. In others, each division becomes a clearly identifiable section. Nevertheless, even in extreme cases the divisions exist.

Before the applications of thermostatic control can be discussed, it is necessary to understand the principles and design of thermostats themselves. This forms the subject of the first part of this book, which ends, at Chapter 6, with an introduction to the general principles and types of control systems. The second part of the book deals with the various applications of thermostatic control, for domestic, commercial and industrial purposes.

1

DIFFERENTIAL-EXPANSION METALLIC THERMOSTATS

ROD-AND-TUBE THERMOSTATS

THE rod-and-tube thermostat consists of a tube, generally of 65/35 brass, enclosing a rod of metal having a lower coefficient of expansion. The rod is attached to the tube at one end, the other end remaining free. Thus for an increase in temperature the brass tube will extend and draw in the inner metal rod. The relative movement between the free ends of the rod and the tube provides the actuating mechanism of this type of thermostat, and, for a rod of given material, will be greater the lower the coefficient of expansion of the rod. In the early experiments with this type of thermostat the rod was made of steel of normal composition but the discovery of Invar, with its extremely low coefficient of expansion, made possible considerably greater relative movements between the free ends of the rod and the tube, and so simplified the thermostat designer's problem of creating a thermostat having a usable thermal movement within reasonable dimensions.

Invar steel, with its 36 per cent nickel content, has low coefficient of expansion at temperatures up to 100°C. Above this temperature the rate of expansion increases, so that whilst Invar offers considerable advantages for use in thermostats measuring temperatures between 0 and 100°C, it can only be of use in applications involving higher temperatures if a non-linear range can be used. Domestic-oven thermostats for both gas and electric applications require a temperature range between 100 and 250°C, for which it is generally found more practical to use a rod with a nickel content of 40 or 42 per cent. These alloys have a somewhat higher coefficient of expansion than the 36 per cent nickel, but maintain a linear performance over a longer temperature band. Other alloys are available under various proprietary names having linear performances up to 400°C. However, as Fig. 1.1 shows, it is a general rule that the longer the temperature band of linear performance the higher is the coefficient of expansion of the alloy.

A linear performance is not always essential. It is most desirable for such items as domestic thermostats where the range knob features as a prominent part of the appliance, and where uniformly-spaced markings contribute to general appearance. In laboratory and commercial thermostats a non-linear performance can easily be accommodated by a correction incorporated into the knob markings, or the use of a non-linear cam instead of a range screw.

The rod-and-tube thermostat has proved itself a reliable and cheap means of providing thermostatic control in the domestic cooking oven over many years, and can provide either modulating or proportional

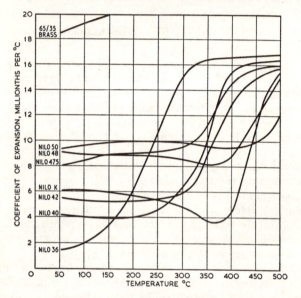

Fig. 1.1 Characteristics of the various controlled expansion alloys produced are declared by manufacturers. The graph shows the characteristics of the Nilo series of controlled expansion alloys produced by Henry Wiggin and Co. Ltd.

action for a gas valve, or (through a suitable linkage) snap-action control of an electric switch.

Rod-and-tube design for gas thermostats

In the design of a rod-and-tube thermostat for a gas control many factors need to be taken into consideration; a simplified explanation follows.

The essential features of a gas thermostat using this type of thermal element are shown in Fig. 1.2. It will be seen that the brass tube and steel rod are joined together at one end and fixed to a gas valve at the other. The method of joining is such that as the expansion of the brass tube takes place, with rise in temperature, its length extends so that the steel rod is withdrawn from the body of the thermostat. The total movement of the end of the steel rod from the body of the valve will be equal to the expansion of the brass tube less the expansion of the steel rod. The gas valve is in hard contact with the rod so that it follows the movement of the rod, a suitable valve seat being incorporated in the body. As expansion takes place the valve closes.

Virtually any length of rod-and-tube thermal assembly can be used for temperature control so long as the valve is arranged and positioned so that it touches the seat and cuts off the gas supply when the rod and

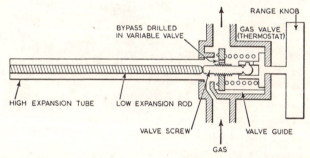

Fig. 1.2. Diagrammatic illustration of the gas rod-and-tube thermostat commonly used on domestic appliances, water heaters and central-heating boilers. (Concentric.)

tube has reached the temperature at which thermostatic control is required. Variation of the length of the thermal assembly determines the rate of movement of the valve towards its seat per degree of temperature. This is an important factor in determining the rate of response of the thermostatic valve and the degree of proportional control it effects over the gas supply. The size of the gas valve will also have an effect so that to determine the correct proportional effect both length of thermal rod-and-tube assembly and valve diameter must be taken into consideration.

In the domestic gas cooker it is normal practice for full gas rate to be maintained up to a point where the temperature in the oven is within some 10°C of the control point, and for the gas rate then to be proportionately reduced to the rate required to maintain the predetermined temperature. If a shorter proportional range than this is used some overshoot may be experienced since the gas input may make the temperature rise at a speed faster than the rod-and-tube thermal assembly can absorb. This lag in reaction may cause the temperature in the oven to rise above the desired value before it is brought under control. If, on the other hand, the proportional effect of the thermostat is such that the gas rate is reduced when the oven temperature is some 20 or 30°C below the control point, sluggish heating up and recovery conditions may be experienced.

It has been found that satisfactory conditions for the control of the domestic gas oven can usually be achieved by using a 254 or 305 mm thermal assembly operating a gas valve of approximately 13 mm diameter.

The use of rod-and-tube thermostats in gas cookers is discussed in greater detail in Chapter 14.

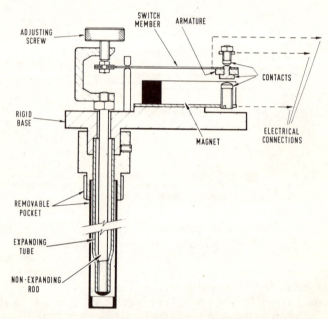

Fig. 1.3. Immersion thermostat using a rod-and-tube thermal element.
(Satchwell Controls Ltd.)

Rod-and-tube thermostats for electric switches

Fundamentally, the rod-and-tube type of thermal element as applied to operate electric switches does not differ from the gas rod-and-tube device already discussed. The main difference is that we are now mainly concerned with the conversion of the slow and somewhat small thermal movement into the positive snap action of a switch. It is, therefore, often necessary to magnify the thermal movement by means of levers and to bring the last lever, carrying a moving contact, into the sphere of influence of a permanent magnet (see Fig. 1.3). A lever action is thus introduced so that the final closing and opening action of the contact is under the control of the magnet and spring leaf which partly overrides the thermal movement obtained from the rod and tube.

In this arrangement the travel of the moving contact towards the fixed contact is under the influence of the thermal movement until it approaches the last stages, when the lever carrying the moving contact comes into the field of the magnet which snatches it forward and so provides a quick snap-action closure. On a rise in temperature, the rod of the thermal element recedes from the switch and attempts to move the contact to the open position. Initially however, the spring resists the action of the thermal element and allows the switch to remain in the closed position under the influence of the magnet. However, as the rod recedes farther it will eventually overcome the influence of the spring and magnet and cause the switch to break circuit. As soon as the magnetic field is broken its force will diminish rapidly as the contacts move apart, and so a relatively quick or snap-action break will be provided.

Other switch designs dispense with the magnet and use instead some form of top dead-centre device with a spring movement ensuring that the switch moves quickly from one extreme to the other.

Either of these devices can be made adjustable so as to control the differential, i.e. the temperature change that takes place between the points of switch opening and closing. The limitation on such an adjustment is determined by the contact pressure and the rate of closing required in relation to the electrical load being carried. Generally the greater the electrical load the greater the need for positive snap-action and good contact pressure, and it is sometimes necessary to provide a substantial magnetic pull. This in turn causes the instrument differential to be somewhat coarse. Where low currents are being controlled the magnetic pull can be considerably reduced with the result that the differential can be made relatively small. Thus it is a general rule that fine differential thermostats of this type are restricted in their maximum loading.

The above description covers a snap-action device of a magnetic type, but there are several others ways of achieving the same result by utilizing purely spring action. Some of these are discussed in Chapter 5.

BI-METAL THERMOSTATS

The term bi-metal is generally applied to a composite metal component consisting of two or more layers of metal bonded together, each

metal having a different coefficient of expansion. Such a metal component changes its shape when subjected to a change in temperature,
the amount of change depending on: (1) the difference between the
coefficient of expansion of the metals bonded together; (2) their
relative thicknesses; (3) their moduli of elasticity; (4) the temperature
change.

Consider two strips of metal of the same length (Fig. 1.4(a)), one strip
A having a coefficient of expansion of 2×10^{-6} and the other B of
4×10^{-6}. If these strips are both 50·8 mm long at 100°C their lengths at
200°C will be 50·80010 mm and 50·80020 mm for A and B respectively.
If instead of being separate these two strips are bonded rigidly together,
and it is arranged that the composite strip is linear at 100°C, when it is
heated to 200°C considerable stresses will be set up in both metals.
Metal A will be subjected to tension by the action of metal B in its
attempt to expand to a greater length than A. In turn, metal B, having
been restrained by metal A, will be in a state of compression. This con-

(a)   (b)

Fig. 1.4. Principle of bi-metal construction. (a) Two separate pieces of metal of different
coefficients of expansion would on their own be free to expand at different rates per
degree rise in temperature. (b) If the same two metals are welded together and sub-
jected to a temperature rise the bonded metal will warp throughout its length.

dition will exist throughout the whole length of the strip, and, as a
result, bending will take place uniformly throughout its length (Fig.
1.4(b)).

Early bi-metal strips consisted mainly of combinations of brass and
steel. These were followed later by brass and Invar (which has already
been mentioned in connexion with the rod-and-tube thermostat). It was
the discovery of Invar, with its low coefficient of expansion, that established the bi-metal as a reliable thermal element suitable for mass production. Today a whole range of high-nickel steel alloys is available for
use as the low expansion component of bi-metals. Further progress in
this field led to the production of other alloys of the iron-chrome-cobalt
family which, in addition to their low-expansion qualities, offer various
degrees of corrosion resistance.

For the high-expansion component, considerable progress has been
made since the early dependence on brass. Nickel-iron alloys having
lower nickel contents than the low-expansion metals and with chrome
and manganese additions are in increasing use. One of the latest of these
is the high-manganese copper-nickel alloy with a coefficient of expan-

sion of nearly 30×10^{-6} per degree Centigrade. This very high rate of expansion has produced a range of metals 50 per cent more active than the average used in bi-metals. From these and other metal alloys a very wide range of bi-metals is available to cover the various properties that may be needed for various applications. Important properties include: (1) electrical resistivity, for applications where the bi-metal itself will carry current; (2) resistance to corrosion, for use in water or other conditions likely to attack metal; (3) thermal activity, i.e. maximum or minimum movement required for any given change in temperature; (4) range of maximum sensitivity (see Fig. 1.5).

Metals are generally selected so that their maximum rate of expansion occurs over the temperature range for which the bi-metal is required to operate.

Manufacturers of bi-metals declare most of this information in their catalogues and lists.

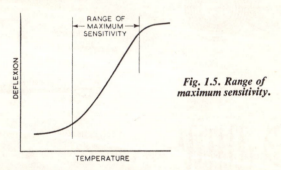

Fig. 1.5. Range of maximum sensitivity.

Thermal activity

This is declared as a strip deflexion constant, which is the deflexion of a strip of unit length and unit thickness for each degree Centigrade rise in temperature. The constant is measured over the linear part of the temperature-deflexion curve. The bi-metals most commonly used have deflexion constants between about 4×10^{-6} and 20×10^{-6}, but for special applications bi-metals with constants above or below this range are available.

Range of maximum sensitivity

Whilst any particular bi-metal may be capable of working over a wide range of temperatures there will be a band within this range where its deflexion rate is highest. This is known as the maximum-sensitivity range, and it is in this particular band that the deflexion constant referred to above is taken and declared (Fig. 1.5).

Design considerations

In the application of thermal metal to thermostat problems the flexing of a beam or other shape can be utilized in three ways.

Free movement. Here the metal is allowed to move with no opposition, so that for any change in temperature to which the metal is subjected it will move through its maximum theoretical deflexion.

Force. In the other extreme case the metal is completely restrained and is incapable of moving in any direction. In these conditions a rise in temperature will generate within the metal a restrained force.

Movement and force. In this combination of the other ways the properties of the bi-metal are used to produce both movement and force. For example, the metal can be arranged to deflect so far and then

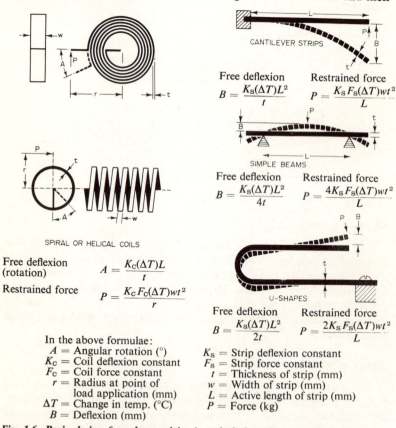

CANTILEVER STRIPS

Free deflexion

$$B = \frac{K_\text{S}(\Delta T)L^2}{t}$$

Restrained force

$$P = \frac{K_\text{S} F_\text{S}(\Delta T)wt^2}{L}$$

SIMPLE BEAMS

Free deflexion

$$B = \frac{K_\text{S}(\Delta T)L^2}{4t}$$

Restrained force

$$P = \frac{4K_\text{S} F_\text{S}(\Delta T)wt^2}{L}$$

SPIRAL OR HELICAL COILS

Free deflexion (rotation)

$$A = \frac{K_\text{C}(\Delta T)L}{t}$$

Restrained force

$$P = \frac{K_\text{C} F_\text{C}(\Delta T)wt^2}{r}$$

U-SHAPES

Free deflexion

$$B = \frac{K_\text{S}(\Delta T)L^2}{2t}$$

Restrained force

$$P = \frac{2K_\text{S} F_\text{S}(\Delta T)wt^2}{L}$$

In the above formulae:

A = Angular rotation (°)
K_C = Coil deflexion constant
F_C = Coil force constant
r = Radius at point of load application (mm)
ΔT = Change in temp. (°C)
B = Deflexion (mm)

K_S = Strip deflexion constant
F_S = Strip force constant
t = Thickness of strip (mm)
w = Width of strip (mm)
L = Active length of strip (mm)
P = Force (kg)

Fig. 1.6. Basic design formulae used in the calculation of movement or force that can be obtained from various forms of bi-metal. (Telcon Metals Ltd.)

be restrained so that it produces a force capable of moving a transmission member of a thermostat and ultimately causing some work to be
done. This is by far the most useful adaption of bi-metallic properties
and is the underlying principle of most bi-metal thermostatic devices.

These points are analysed and determined at an early stage in any
thermostat design problem as they considerably influence the further
steps to be taken.

The main points to be taken into account in designing a bi-metal
thermostat are: (1) temperature range over which thermostatic control
is required, and the maximum temperature to which the thermal metal
is likely to be subjected; (2) force and movement required; (3) working
conditions and their effect on the metals to be chosen; (4) space available—establishing the overall space available for the bi-metal helps to
decide the shape and total effective length that will be required.

Normally, it is found that the metal offering the most movement
possible, and yet complying with all the other factors above, is the final
selection for any one application. With these last points decided it is
possible, using manufacturer's formulae, to obtain an approximate idea
of the metal, type and thickness that will be required to produce the
movement and force needed over the desired temperature range.
Fig. 1.6 shows some standard forms of bi-metal strips in common
use and typical equations used to provide the design information
required. Table 1 shows the deflexion constants for several standard

Table 1. Deflexion constants
(in millimetres, kilograms and °C)

Bi-metal type	Strip deflexion constant (K_S)	Strip force constant (F_S)	Coil deflexion constant (K_C)	Coil force constant (F_C)	Electrical resistivity ($\Omega\ mm^{-2}\ m^{-1}$)
200	$19{\cdot}9 \times 10^{-6}$	$3{\cdot}4 \times 10^3$	$23{\cdot}5 \times 10^{-4}$	$20{\cdot}1$	$1{\cdot}09$
160	$15{\cdot}4 \times 10^{-6}$	$4{\cdot}7 \times 10^3$	$20{\cdot}8 \times 10^{-4}$	$25{\cdot}2$	$0{\cdot}78$
140	$14{\cdot}0 \times 10^{-6}$	$4{\cdot}7 \times 10^3$	$18{\cdot}9 \times 10^{-4}$	$25{\cdot}2$	$0{\cdot}76$
131	$13{\cdot}3 \times 10^{-6}$	$3{\cdot}25 \times 10^3$	$18{\cdot}0 \times 10^{-4}$	$18{\cdot}3$	$0{\cdot}18$
400	$12{\cdot}0 \times 10^{-6}$	$4{\cdot}7 \times 10^3$	$16{\cdot}2 \times 10^{-4}$	$25{\cdot}2$	$0{\cdot}70$
15	$9{\cdot}5 \times 10^{-6}$	$4{\cdot}7 \times 10^3$	$12{\cdot}8 \times 10^{-4}$	$25{\cdot}2$	$0{\cdot}17$
188	$9{\cdot}0 \times 10^{-6}$	$4{\cdot}35 \times 10^3$	$12{\cdot}2 \times 10^{-4}$	$23{\cdot}2$	$0{\cdot}93$
11	$8{\cdot}4 \times 10^{-6}$	$4{\cdot}35 \times 10^3$	$11{\cdot}3 \times 10^{-4}$	$23{\cdot}2$	$0{\cdot}20$
75	$6{\cdot}8 \times 10^{-6}$	$4{\cdot}7 \times 10^3$	$9{\cdot}2 \times 10^{-4}$	$25{\cdot}2$	$0{\cdot}57$
41	$4{\cdot}1 \times 10^{-6}$	$4{\cdot}7 \times 10^3$	$5{\cdot}5 \times 10^{-4}$	$25{\cdot}2$	$0{\cdot}16$
38	$3{\cdot}8 \times 10^{-6}$	$4{\cdot}7 \times 10^3$	$5{\cdot}1 \times 10^{-4}$	$25{\cdot}2$	$0{\cdot}56$

Note—Bi-metals can be used to produce deflexion or force, or a combination of
both of these. Tables of this type show the deflexion or force that can be developed
per degree centigrade rise in temperature for various bi-metals. (Telcon Metals Ltd.)

bi-metals, provided by the manufacturer for use with the formulae given under Fig. 1.6. In view of the very large number of bi-metal combinations that now exist, and of the fresh ones that become available as further research work is carried out, it is inadvisable at an early stage to make specific recommendations for the use of any one type of bi-metal for any particular application. The designer or other person interested in the type of bi-metal employed must ensure that he is equipped with the latest information from manufacturers so that he can select the best of possibly three or four metals capable of doing the work in hand. Although the formulae can eliminate a considerable amount of trial and error work it must also be emphasized that in the end a considerable amount of experimental work under actual working conditions should take place before a final bi-metal choice is made. So much depends on the local conditions under which the metal must operate that exact calculation by formulae is impossible.

Thermal metal is available from the manufacturers in strip or coil form for fabrication into its final shape as a thermal element. In its original strip form it will have been annealed as part of the manufacturing process; but any work subsequently carried out on the metal to produce its final shape for use in a particular thermal device must have some effect on its properties. Operations such as forming, welding, riveting or clinching, used in fixing the thermal metal into position, introduce stresses within the metal, some of which will be fairly evenly spread throughout the length, whilst others will be localized round points of cold working. These will be redistributed as soon as the metal is heated, and in some cases will result in the metal assuming a slightly different shape on cooling. For this reason the precaution should normally be taken of ensuring that the metal is heated after all work on it has been completed, but before it is calibrated or zeroed in the complete thermal device. This process is normally restricted to holding the bi-metal at a temperature of 30°C in excess of the maximum temperature likely to be experienced in service for about an hour, after which it is allowed to cool naturally. If the bi-metal is then calibrated within its thermal device no further drift should normally be expected.

Typical bi-metal applications

The flexibility of possible ways in which bi-metals may be used can be seen from Figs. 1.7 and 1.8. Both show coil-type elements. Fig. 1.7 shows an element designed for use in a small temperature indicator where a comparatively thin coil of bi-metal unwinds or winds up in relation to temperature and moves a pointer over a dial. Here the work is negligible, against zero resistance, and apart from moisture in the air the device has no adverse operating conditions to contend with. Fig. 1.8 shows a coil of bi-metal with a similar operating principle to Fig. 1.7, but where the movement required, work to be done, and operating conditions are much more severe. The coil is responsible for operating a three-way mixing valve on a domestic hot-water circuit. The bi-metal is immersed in the water circuit and is subject to attack from scale and other foreign matter; but experience has shown that it is quite capable of operating

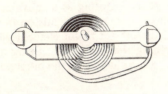

Fig. 1.7. Coil bi-metal used for thermometer indicator.

Fig. 1.8. Coiled bi-metal for use with hot water mixing valve.

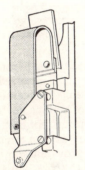

Fig. 1.9. Thermostat bi-metal.

Fig. 1.10. Immersion bi-metal thermostat.

under these conditions over long periods. Fig. 1.9 shows a typical 'U' shaped element for a room thermostat, in which the bi-metal has to provide a combination of force and movement. Some of the movement must be restrained to provide sufficient force to operate an electric switch suitable for controlling loadings of approximately 2 kW. Fig. 1.10 shows a coiled bi-metal for use in an immersion heater where maximum movement is required but where the thermal element has to be contained within a tube of reasonably small diameter. The coil is a natural choice for this type of application. In the type illustrated, the initial position of the bi-metal element in relation to the fixed contact can be adjusted by the range knob to control the travel required of the moving contact before it completes the electric circuit.

Reversed bi-metal

In addition to the standard form of bi-metallic strip used in the applications described above, it is sometimes found useful to employ a strip of bi-metal with some of its length reversed. In a thermostat element consisting of a single cantilever strip operating as a movable contact, making and breaking the contacts is slow and depends on the rate of temperature change. A reversed thermostat metal element as shown in Fig. 1.11 can assist in overcoming this limitation, giving a positive make and break of the contacts. In the illustration the bi-metal

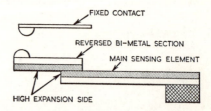

Fig. 1.11. Reverse bi-metal used on cantilever type thermal switch.

carries the electric current. With a fall in temperature the main sensing element travels towards the fixed contact but the small reversed bi-metal section at the end moves in the opposite direction (but because of its short length it has little effect on the final point of making contact). As soon as contact is made, current begins to flow through the bi-metal, increasing its temperature, so that the small reversed section attempts to move rapidly in the direction of the fixed contact, providing considerable contact pressure and eliminating any tendency to spark. On a rise in ambient temperature the main sensing strip will tend to move away from the fixed contact, but electrical continuity will be retained until the restraining force of the small reversed section no longer exists. As soon as the sensitive element is moved so far that it breaks circuit, current ceases to flow through the bi-metal and the small reversed length moves rapidly away from the fixed contact, again eliminating the possibility of prolonged sparking. By choosing bi-metal with suitable electrical resistivity and movement per degree temperature change a combination of sensing strip and reversed-action strip can be designed to make full use of this action without destroying, or unnecessarily extending, the differential of the thermostat.

The same basic principle is useful in the control of the movement of bi-metals for such applications as electric irons, where initial over-shoot can be prevented. Similarly it is frequently found useful in flame-failure devices and excess-temperature devices where the bi-metal is required to open a gas valve as soon as a flame has been established and where it is expected to close the valve as quickly as possible should the flame be extinguished. To achieve the first aim, the bi-metal will be required to open the gas valve when it is being heated by the gas flame but is still relatively cool. On the other hand, it is expected to shut off the gas as soon as the flame has been extinguished even though at this moment the bi-metal may be at a relatively high temperature. These two conditions can be met by introducing reversed bi-metal so that on the temperature run up an initial movement is obtained on the first rise in temperature but subsequent excess movement caused by further rise in temperature is overcome by the reversed section of the bi-metal strip. On a fall in temperature caused by flame failure the first movement of the bi-metal is immediately conveyed to the valve and the gas is cut off even though the bi-metal as a whole may still be considerably higher in temperature than it was at the instant when the gas valve was opened.

Bi-metallic snap-action switches

An interesting use of bi-metal is found in the Otter patented principle in which thermal sensitivity and snap action are combined by the design of the bi-metal blade (Fig. 1.12(a)). Maximum thermal sensitivity is

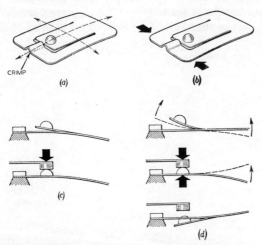

Fig. 1.12. Various stages in the action of a snap-action bi-metal switch. (Otter Controls Ltd.)

obtained by using a large area of thin bi-metal and making use of the transverse as well as the longitudinal flexion. Thin bi-metal is used because of its large surface area in relation to thermal mass, and because thermal flexion of bi-metals is inversely proportional to thickness.

The blade has three 'legs' and has a built-in snap-action because the crimp draws the outer legs inwards putting an over top-dead-centre stress in the bi-metal blade (Fig. 1.12(b)). The centre leg is then forced downwards when assembled into a thermostat to give a high contact pressure (Fig. 1.12(c)). Fig. 1.12(d) illustrates one of the most important attributes a thermostat can have, i.e. highest contact pressure at the point of operation. As the bi-metal flexes upwards, so does the centre leg which carries the moving contact, but it is restrained by the fixed contact. These thermostats press their contacts together, and open them with a snap, which ensures minimum contact wear and dispenses with the need for separate snap-action mechanisms.

In addition to their use in thermostats these thermal switches may also be arranged for the current-temperature protection of electric motors and as current circuit breakers.

In addition to the bi-snap action device mentioned above there are several other means of forming the bi-metal in such a way that it produces a snap action. Generally these take the form of discs of metal 'domed' or corrugated in such a way that changes in temperature gradually build up stresses within its shape but where actual movement is resisted. Further changes in temperature cause the stress to be overcome and the metal shape goes through an instantaneous change, the Klixon series of thermostats by Salford are typical of this device.

2

CAPSULE-TYPE
ELEMENTS

WHEN subjected to a rise in temperature, liquids first expand and then, as they reach their boiling points, evaporate. If the vapour driven off in the boiling process is enclosed within a capsule it will be capable of creating a pressure in relation to temperature. If this saturated vapour is heated still further it will be converted to a dry gas which again is capable of expanding with further increases in temperature. For each liquid, vapour, or gas stage definite conditions of expansion or pressure prevail for any given temperature. Therefore, if a liquid is enclosed within an expanding or expandable capsule, which is then hermetically sealed, there will be a definite temperature-movement relationship for any given temperature.

This principle is the basis of many thermal devices used in commercial and domestic thermostats. For the most part, the capsule takes the form of a flexible diaphragm, bellows, or Bourdon tube. In one type of instrument the liquid or vapour is enclosed in a separate temperature-sensing phial immersed in the controlled medium and connected to the capsule by a length of capillary tubing whose volume is small in relation to that of the capsule. On the other hand, when the capsule is a bellows or a diaphragm, it is often more convenient to make this self-contained, i.e. to dispense with the separate sensing phial and capillary and to arrange for the capsule itself, charged with the appropriate liquid or vapour, to sense the temperature changes to be controlled.

BOURDON TUBES

Of the three basic designs of capsules in use for the measurement of pressure change and liquid expansion the Bourdon tube is undoubtedly the oldest. It originates from Bourdon's work in the middle of the nineteenth century when the universal use of steam as a source of power created a demand for methods of measuring pressure. Since that time the Bourdon tube has been successfully applied to numerous types of instruments for measuring, indicating, and controlling pressure; and it has been usefully applied in thermostatic-control systems in which the temperature-sensing device operates on a temperature-pressure relationship, or temperature-liquid-expansion relationship. Generally its size prevents it from being used on anything other than the larger industrial instruments where operation on a fine differential at high pressures is required.

It generally consists of a tube of oval section bent along its length into an arc (Fig. 2.1). One end is sealed and the other end is connected through a boss to the source of pressure change. A rise in pressure within the tube tends to convert the oval tube into a cylindrical one and, at the same time, to straighten out the arc, causing the tip to move outwards. The progress of this straightening process is extremely complex. Many theories have been suggested in attempts to provide a formula for calculating the movement obtained by pressure change in the Bourdon tube. For our purpose it is sufficient to assume that the movement of the tip is substantially linear in relation to the change in pressure.

Construction

The majority of Bourdon tubes in commercial use are produced by flattening a cylindrical tube to the required oval section. In some cases the walls are left well apart, whilst in others the tube is flattened to a point where its volumetric capacity is virtually nil. The flattened tube is then bent either into a circular, subtending some 260–280°, angular, or, when greater movement is required, and where the tube is considerably longer in its initial state, it is wound into a spiral or helical form. The open end of the tube is brazed to a substantial boss suitably drilled and threaded to accept a connexion from the pressure system. The same boss also forms the base for mounting any linkage required to multiply the movement obtained from the tube. The sealed end of the tube is connected to this linkage through a pinion which engages a quadrant pivoted on to the centre mechanism. In this way the linear movement of the tip of the tube is converted to a rotary movement capable of driving a pointer or, in the case of a thermostat, operating some switch or valve.

According to the working conditions and the pressure range over which the tube has to operate, the materials employed for the manufacture of the tube may be steel, brass, phosphor-bronze, beryllium-copper or K. Monel metal, the first two being the most used. The Bourdon tube, being a highly sensitive device, needs some protection against excessive vibration and rapid pressure pulses. The latter can be overcome by introducing a throttle in the pressure connexion so that rapid changes in pressure are damped out and the Bourdon tube is subjected only to the mean of the oscillations. Vibration can be damped

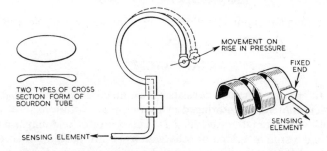

TWO TYPES OF CROSS
SECTION FORM OF
BOURDON TUBE

SENSING ELEMENT

MOVEMENT ON
RISE IN PRESSURE

FIXED
END

SENSING
ELEMENT

Fig. 2.1(a). Typical forms of Bourdon tube used in thermostat construction.

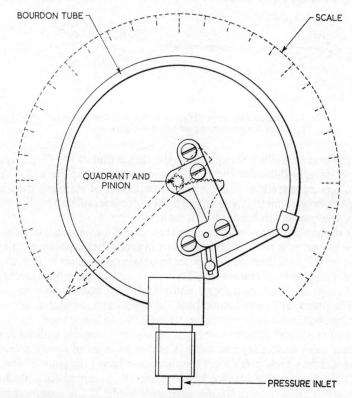

BOURDON TUBE

SCALE

QUADRANT AND
PINION

PRESSURE INLET

Fig. 2.1(b). Bourdon tube driving a pointer to give indication of temperature.

out by coiling the capillary connexion between the sensing element and the Bourdon-tube boss.

DIAPHRAGMS

The diaphragm generally consists of two metal discs welded together at their circumference and ringed in a series of convolutions to provide flexibility with strength (see Fig. 2.2). The centre of one diaphragm is drilled and a stud is welded or otherwise fixed into position to accept a connexion between the diaphragm and a bulb containing the thermally sensitive substance. The metal chosen for the diaphragm depends on the work it is to do, the type of liquid being used, and whether the diaphragm is expected to be moved by hydraulic pressure caused by thermal expansion of the liquid or by vapour pressure created by boiling liquid.

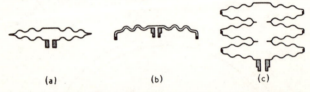

(a)   (b)   (c)

Fig. 2.2. Typical diaphragm forms used in temperature-pressure and temperature-liquid expansion systems.

In the case of hydraulic expansion, the spring rate of the diaphragm will not greatly influence the performance of the complete unit. The diaphragm is, therefore, a fairly rigid component of stainless steel or phosphor-bronze metals, which, with their considerable strength, will satisfy the particular needs of this unit.

For vapour-pressure systems where the thermal movement is obtained by low pressures it is an advantage to arrange for the diaphragm spring rate to be comparatively low, in order to obtain reasonably large movement for a given temperature rise. To meet these various requirements a whole range of diaphragm forms and metals have come into common use. The only real limitations on both the design and the metals are the resistance of the metal to corrosion with the liquid being used, and the ability of the metal to deflect over the required movement without fear of either work hardening and early failure or changes of spring rate or permanent set which would alter the characteristics of the system. These points, in fact, in the absence of any generally accepted formula for diaphragm design, constitute the designer's guide in the development of a new diaphragm. They have been backed by extended life tests for most of the diaphragms in general use.

Movement

One of the limiting factors of a diaphragm is the restricted working movement it can produce. All the movement obtained by deflecting the diaphragm must come from the two metal discs whose deflexion must be well within their moduli of elasticity. A 31·75 mm stainless steel diaphragm of 0·250 mm wall thickness will provide a useful working movement of some 0·760 mm. A 50·0 mm phosphor-bronze diaphragm will move some 1·8 mm. The total movement of the two discs is, therefore, quite small, so that where work is needed with a small change in temperature this movement must either be multiplied mechanically, or the diaphragm diameter must be so large that it becomes impracticable for general applications. A series of diaphragms is, therefore, sometimes used, connected together so that the complete assembly has the total movement of the separate diaphragms (Fig. 2.2(c)).

BELLOWS

The bellows is a development of the diaphragm and provides a convenient way of producing a movement equivalent to that obtained from several diaphragms in series (Fig. 2.3). The metal used in bellows manufacture must be a ductile alloy capable of being drawn and worked without fear of splitting or becoming porous. For the most part a brass is chosen, consisting of 80 per cent copper and 20 per cent zinc. This alloy has excellent deep-drawing qualities, and after cold working has good spring properties which are needed in most thermal systems. Other metals can be used where special conditions call for their particular properties. Phosphor-bronze of 95 per cent copper and 5 per cent tin is frequently used where resistance to corrosion is important. In addition phosphor-bronze produces a slightly more stable bellows than one made of brass. Alumbro metal, nickel-silver, aluminium-bronze, cupro-nickel, beryllium-copper, and stainless steel are also used for bellows where their particular properties are required. However, apart from phosphor-bronze, metals other than 80/20 brass tend to be stiffer when cold-worked and, therefore, although they have the advantage of higher corrosion resistance, they have the disadvantage that they generally provide bellows having higher spring rates.

CAPSULE SELECTION

In the design of a thermal system utilizing a capsule (diaphragm, bellows or Bourdon tube) several factors need consideration before the capsule can be selected. In a hydraulic system, where the thermal movement is derived from the expansion of a liquid, the movement of

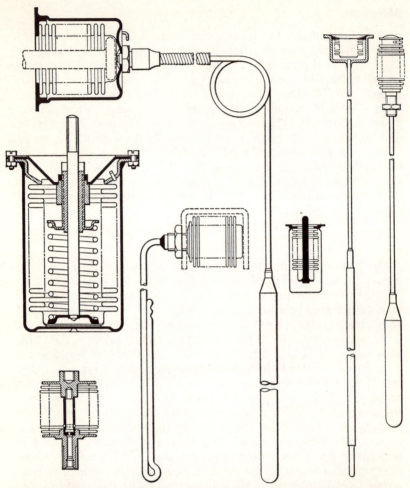

Fig. 2.3. Bellows operated thermal systems. The two systems on the right-hand side, having comparatively small bellows, are of use in hydraulic expansion systems. The other designs are typical of vapour-pressure systems. (Drayton Controls Ltd.)

the capsule is fixed by the quantity and properties of the liquid and will not be affected by the spring rate of the capsule or the opposition created by external forces, such as springs or water or steam pressure. Therefore, the calculation will only be concerned with: (1) the total volume of liquid in the system likely to be subjected to changes in temperature; (2) the maximum expansion likely to be experienced; and

(3) the final selection of a bellows or diaphragm or Bourdon tube that can expand and thus increase its volumetric capacity to accommodate the expansion of the liquid without over-taxing its mechanical stability.

Very often these factors will be considered in the reverse order in that it will first be decided that a certain capsule movement is required to do a certain amount of work to move a valve or switch, and the first step will therefore be to select a capsule capable of giving the movement required. The change in volumetric capacity of the capsule will then be noted and finally the thermal sensing bulb will be sized so that it will have the correct amount of liquid within it to expand and in its expansion provide the amount of movement required of the capsule.

In either method of calculation the selection of the right bellows, diaphragm, or Bourdon tube, is comparatively simple, and comprises only the choice of unit that comes within the manufacturers' tables of allowable movement.

The choice of a capsule for a vapour-pressure or gas system is somewhat more complicated. Here the movement of the bellows, or other capsule, corresponds to the change in pressure created by a change in temperature. It has been previously stated that the finished bellows is capable of acting as a spring; therefore, its own spring rate, as well as the change in pressure in the thermal system, will affect the movement.

Whilst the bellows acts as a spring in its final state it does not follow the exact formula of spring design. A bellows will be much more restricted in its degree of compressibility than a spring of comparable size. This is in the main due to the fact that, whereas a spring derives its qualities from the material chosen for its construction, the bellows material in its initial state has little spring quality, the limited final amount being produced by the work hardening it receives during the bellows forming. Furthermore, whereas it is possible for a spring to work in extension, a bellows is rarely able to do this for any length of time and is nearly always arranged to work between its normal free length and the maximum compression declared by the manufacturer.

Each manufacturer provides a list of bellows available with details of certain standard features of each. Such a list, an example of which is given in Table 2, assists designers in the selection of the correct bellows for a thermal system by providing, amongst other things, details of the maximum movement that can be expected from a bellows of a given length and wall thickness, and the maximum unbalanced pressure to which the bellows can be subjected. At least one manufacturer of bellows gives these figures in relation to the useful life that can be expected from any one unit. This information is important since, while a bellows worked at its maximum movement and maximum out-of-balance pressure will have a certain life, its life will be considerably longer if

Table 2. Range of single-ply brass bellows

Ref. no.	Convolution outside diameter		Convolution root diameter		Wall thickness		Full number of convolutions	Effective area		Maximum deflexion in compression				Spring rate				Flexibility per convolution		Maximum internal pressure		Maximum external pressure	
										One convolution		Full number of convolutions		One convolution		Full number of convolutions							
	in	mm	in	mm	in	mm		in²	cm²	in	mm	in	mm	lb/in kg/mm		lb/in kg/mm		lbf/in²	mm/bar	lbf/in²	bar	lbf/in²	bar
05104*	5/16	7·9	13/64	5·2	0·004	0·10	21	0·06	0·4	0·0038	0·09	0·080	2·0	1390 24·8		66·5 1·19		0·00004	0·015	700 48·3		770 53·1	
06145*	3/8	9·5	1/4	6·4	0·0045	0·11	23	0·074	0·5	0·0065	0·17	0·150	3·8	990 17·7		43 0·77		0·00007	0·026	280 19·3		300 20·7	
07104*	15/32	11·9	5/16	7·9	0·004	0·10	17	0·12	0·8	0·0092	0·23	0·156	4·0	612 10·9		36 0·64		0·00019	0·070	350 24·1		385 26·5	
09104*	9/16	14·3	3/8	9·5	0·004	0·10	26	0·164	1·0	0·012	0·30	0·312	7·9	676 12·1		26 0·46		0·00025	0·092	210 14·5		230 15·9	
09105*	9/16	14·3	3/8	9·5	0·005	0·13	25	0·164	1·0	0·011	0·28	0·265	6·7	1092 19·5		44 0·79		0·00016	0·059	290 20·0		310 21·4	
09107*	9/16	14·3	3/8	9·5	0·007	0·18	26	0·164	1·0	0·008	0·20	0·210	5·3	3250 58·0		125 2·23		0·00008	0·029	565 39·0		625 43·1	
12104*	3/4	19·1	1/2	12·7	0·004	0·10	26	0·31	2·0	0·0193	0·49	0·500	12·7	334 6·0		13 0·23		0·00093	0·343	100 6·9		110 7·6	
12105*	3/4	19·1	1/2	12·7	0·005	0·13	26	0·31	2·0	0·018	0·46	0·468	11·9	615 11·0		24 0·43		0·00050	0·184	130 9·0		140 9·7	
12106*	3/4	19·1	1/2	12·7	0·006	0·15	17	0·31	2·0	0·0128	0·33	0·218	5·5	1088 19·4		64 1·14		0·00029	0·107	180 12·4		200 13·8	
15104*	15/16	23·8	5/8	15·9	0·004	0·10	13	0·48	3·1	0·024	0·61	0·312	7·9	208 3·7		16 0·28		0·0023	0·847	70 4·8		77 5·3	
15106	15/16	23·8	5/8	15·9	0·006	0·15	13	0·48	3·1	0·0193	0·49	0·250	6·4	546 9·8		42 0·75		0·00088	0·324	200 13·8		220 15·2	
18104*	1 1/8	28·6	3/4	19·1	0·004	0·10	11	0·69	4·5	0·0283	0·72	0·312	7·9	154 2·8		14 0·25		0·0045	1·658	55 3·8		60 4·1	
18105*	1 1/8	28·6	3/4	19·1	0·005	0·13	11	0·69	4·5	0·027	0·69	0·297	7·5	264 4·7		24 0·43		0·0026	0·958	80 5·5		88 6·1	
18106* {	1 1/8	28·6	3/4	19·1	0·006	0·15	11	0·69	4·5	0·0256	0·65	0·281	7·1	517 9·2		47 0·84		0·0013	0·479	150 10·3		165 11·4	
18106* }	1 1/8	28·6	3/4	19·1	0·006	0·15	24	0·69	4·5	0·0256	0·65	0·615	15·6	517 9·2		22 0·39		0·0013	0·479	150 10·3		165 11·4	
18107* {	1 1/8	28·6	3/4	19·1	0·007	0·18	11	0·69	4·5	0·0227	0·58	0·250	6·4	704 12·6		64 1·14		0·001	0·368	200 13·8		220 15·2	
18107* }	1 1/8	28·6	3/4	19·1	0·007	0·18	24	0·69	4·5	0·0227	0·58	0·545	13·8	704 12·6		31 0·55		0·001	0·368	200 13·8		220 15·2	
20106	1 1/4	31·8	7/8	22·2	0·006	0·15	11	0·885	5·7	0·0313	0·80	0·343	8·7	528 9·4		48 0·86		0·0017	0·626	120 8·3		130 9·0	

* Also available in phosphor bronze.

(Drayton, Hydroflex Ltd.)

both the movement of the bellows and the out-of-balance pressure is considerably less than the manufacturer's declared maximum. For this reason, most designers select a bellows with a movement and pressure performance somewhat higher than the highest to which it is likely to be subjected in normal life. The degree of this safety factor depends on the number of operations the bellows is expected to provide during a definite period of time.

In the case of a domestic refrigerator thermostat, which would normally cycle several times per hour during its useful life, this safety factor is well worth applying, some increase in the dimensions of the bellows over the theoretically required size for the movement expected of it being an advantage. In the case of an excess-temperature switch which must operate only when conditions are abnormal, however, there is little point in increasing the bellows size in order to give it, say, a life of 10,000 cycles.

Most bellows manufacturers give bellows life in terms of numbers of cycles for given operating conditions, and designers of thermal movements generally select a component well within the maximum working limits of their particular application.

3

LIQUID-EXPANSION AND VAPOUR-PRESSURE SYSTEMS

SYSTEMS of the types described in this chapter employ bellows, diaphragms, or Bourdon tubes which are operated by pressure or expansion of a liquid, gas, or vapour.

LIQUID-EXPANSION SYSTEMS

The advantages of a liquid-expansion system are: (1) as the coefficient of expansion of liquids used in thermal systems is reasonably constant over a certain range it is possible to design the system so that the movement of the bellows or diaphragm is linear and bears an exact relationship to the temperature change; (2) since the liquid the capsule contains is virtually incompressible, its movement is negligibly affected by any external force that may oppose it; (3) capsules offer a long temperature range: any temperature range can be produced from the one liquid provided that the range is within the limits of the liquid stability, i.e. is above its freezing point and below its critical temperature or does not include a point where a permanent change takes place. Liquids in use include mercury, toluene and Aroclor. Against these advantages, however, there are several limitations.

Head sensitivity

The whole system is charged with liquid and is, therefore, sensitive to temperature change, whereas we wish the system to react only to changes of temperature to which the sensing element is exposed. For a self-contained bellows, where the whole unit is designed to react to temperature, this presents no problem; but where the system comprises a capsule connected to a sensing bulb by capillary tubing the problem is a real one. For instance, in a domestic oven thermostat the temperature of the capsule may vary between about $-1 \cdot 0°C$ and $149°C$. Several methods are used to minimize this disadvantage, most of them being based on keeping the volume of the liquid in the sensing bulb as large a percentage of the total volume as possible.

Bourdon tubes and diaphragms present less of a problem than bellows, for both have little capacity of their own, and it can be arranged that, at the bottom of the temperature range of the instrument being designed, no liquid is present in the capsule and that the walls are pressed close together. At the top of the range, therefore, the liquid content of the diaphragm or Bourdon tube will still only be equivalent to the total expansion of the liquid in the rest of the system, so that even if this liquid in the capsule is heated by the temperature around it the 'false' expansion obtained will be quite small.

A bellows offers a somewhat different problem for in its collapsed or compressed state at the bottom of the temperature range it will still have a volume of its own. This can be considerably reduced by filling in most of the bellows space with a pressed lining as shown in Fig. 3.1;

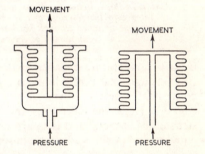

Fig. 3.1. Bellows systems used in hydraulic systems, showing methods adopted to keep the amount of liquid in the region of the bellows at a minimum.

but owing to its construction a small amount of liquid must remain in the bellows and when heated will provide a degree of false movement which must be compensated for or cancelled.

Head compensation, as it is generally termed, provides a means of measuring the temperature surrounding the head of the capsule, anticipating the false movement that this temperature will cause, and correcting this movement before it is conveyed to the valve or switch in the thermostat. The simplest way is to provide a second capsule, self-contained and of the same volume as the thermally-operated one, and charged with the same liquid. Such an arrangement is shown in Fig. 3.2. Here it will be seen that as the temperature around the sensing phial rises bellows A moves forward to operate a valve. If the temperature around bellows A rises the liquid within it will also expand and false movement will be obtained. However, bellows B will be subjected to a similar temperature rise with a similar expansion of the liquid within it, and it will be seen from the diagram that the movement of bellows B is arranged so that it cancels out the false movement obtained by the expansion of the liquid in bellows A. Thus the valve will be moved only

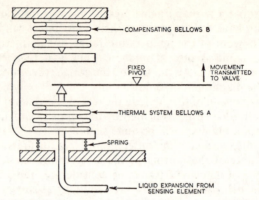

*Fig. 3.2. Method of compensating for head temperature and false move-
ment. In this instance a bellows similar to the thermal bellows is incorporated into the leverage and is charged with the same liquid. False
movement obtained through head temperature is cancelled out by the
movement of compensating bellows B.*

by true temperature changes in the sensing phial. Usually space res-
trictions and cost preclude the use of this arrangement, it being more
convenient to use a bi-metal device to provide the compensation—see
Fig. 3.3. Here, the lever from the bellows to the valve is partly constructed of bi-metal and is so arranged that the metal deflects at a rate
equal and opposite to the false movement of the bellows.

*Fig. 3.3. Alternative method for compen-
sating head temperature utilizing a bi-
metal which will deflect at the same rate
as the false movement obtained from the
bellows.*

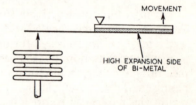

There is one obvious limitation in these methods of compensation,
namely, that of deciding at which point of the temperature range true
head compensation is required. At any temperature above the lowest of
the range the volume of the liquid in bellows A depends not only on the
initial volume of the bellows, but also on the expansion of the liquid
in the phial. On the other hand, the volume in bellows B depends only
on the initial volume. One can only make these two equal at one
temperature, and unless they are equal true compensation is not
achieved. Similarly the alternative method of employing a strip of bi-
metal can only provide for the compensating movement to be equal to
one particular bellows volume/movement.

Head compensation can, therefore, be exact only for one point in the range at a specified head temperature, and it is necessary to decide the temperature at which exact head compensation is required. The fact that over and under compensation is present at above and below this temperature is generally tolerated in commercial instrument work.

Capillary sensitivity

The liquid in the capillary is equally sensitive to temperature change, but since the volume of this liquid is generally small the problem of compensating for the false movement it causes is normally ignored in commercial and domestic thermostat designs. Fine-bore capillary tubes with internal diameters of 0·250 mm or less are commonplace and their use ensures that the percentage of the total liquid in the capillary is small. Where the demand for accuracy is such that compensation must be provided, a length of capillary identical to that used in the thermal system is added to the compensating bellows B in Fig. 3.2.

Thermal movement

The thermal movement obtained from hydraulic expansion is quite powerful and is unaffected by any changing pressures that may oppose it but, unless the sensing bulb is made very large and unwieldy, the movement obtained purely from the expanding liquid will be relatively small. It is quite common for the thermal system of this type designed for domestic applications to provide a total movement of something in the order of 0·75 mm for a temperature change of 150°C. Since the thermostats to which these elements are applied must sense and correct temperature changes of 2·0°C, which corresponds to a thermal movement of 0·0120 mm, it can be seen that conversion of the thermal movement to an operating movement, whether it be a switch controlling the mains voltage or a valve controlling the flow of oil or gas, calls for exact engineering and high accuracy.

Over-run

In both bi-metal and vapour-pressure systems once a specified thermal movement has been achieved and has been converted into work further attempts on the part of the thermal system to move (resulting from further temperature change) can be arrested and, within certain limitations, put to use, for example in increasing the contact pressure in an electric switch without fear of over-stressing the thermal system.

With the hydraulic system, however, this problem is not quite so simple for if, after the thermal system has been made to operate its correcting mechanism, further rises in temperature cause further thermal movement, some positive arrangement must be provided to accommodate this.

This is often referred to as over-run. Over-run arrangements which allow the thermal capsule to continue its movement even though a switch has been operated or a valve has been moved to its fully closed or fully open position are shown in Figs. 3.4 and 3.5.

The first diagram shows a bellows driving a valve, the two being on a common centre line. In normal conditions the bellows will push the valve towards its seat until these two are in contact with each other. Should,

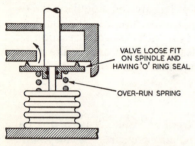

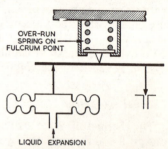

Fig. 3.4. Temperature over-run device used for hydraulic systems operating valves. Excess movement of the bellows once the valve is on its seat is absorbed by the bellows being allowed to continue its forward movement, pushing the valve spindle forward but leaving the loose fitted valve stationary on its seat and in the fully closed position.

Fig. 3.5. Over-run device for liquid expansion system utilizing a sprung fulcrum. Further movement of the thermal element once the valve is on its seat is allowed for by having a sprung fulcrum point.

then, any further temperature rise occur after the valve is seated further bellows movement continues against the pressure of the over-run spring. The centre spindle will continue to pass through the valve and so prevent any damage due to excess pressure built up by the excess thermal movement. In the second example, Fig. 3.5, a diaphragm pushes on a lever the other end of which moves a valve on to its seat. Under normal conditions the diaphragm will move forward until the valve is seated. Should any further temperature rise occur, causing further thermal expansion, the diaphragm will move forward and depress the fulcrum point against the over-run spring. This state of affairs will continue until the excess temperature has been lost when the diaphragm will return to its normal maximum forward movement and allow the fulcrum to resume its normal position.

SOLID/LIQUID PHASE WAX ACTUATORS

In addition to those thermal systems using the normal expansion of a liquid to obtain thermal movement there is a range of actuators which utilize the expansion obtained from hydro-carbon waxes, such as paraffin wax. These produce similar temperature-movement relationships to those of liquids and owing to their robustness and simplicity have proved useful as actuators for such things as damper flaps on solid-fuel boilers.

Being highly viscous, the thermal material is normally used in self-contained elements such as bellows so that changes in viscosity within

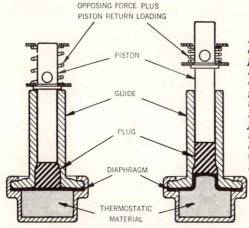

OPPOSING FORCE PLUS
PISTON RETURN LOADING

PISTON

GUIDE

PLUG

DIAPHRAGM

THERMOSTATIC
MATERIAL

Fig. 3.6. Solid/liquid phase wax actuators, general operating principle. Left-hand diagram shows the element at rest in the cold condition. Right-hand diagram shows the heated condition when the thermostatic material has expanded and has forced a synthetic moulded rubber plug into the reduced diameter in the piston guide. (Vernatherm, American-Standard.)

the temperature range over which the element has been designed to operate do not materially affect the performance. The temperature-movement relationship for these solid thermal elements is linear, but the movement per degree temperature change is relatively small. Hence some mechanical multiplication has to be employed to obtain a useful movement at any specified set-point.

An interesting extension of this basic principle has been developed in the U.S.A. by Vernatherm. A range of waxes has been developed which expand considerably as they are changed by temperature from the solid to the liquid state. By enclosing this thermal material in a robust container, considerable improvement is obtained in the temperature-movement-force characteristics normally obtainable from other types of hydraulic expansion movements. Fig. 3.6 shows the arrangement of the sensing element both in its cool and heated states. The thermostatic

material expands with temperature increase to force a moulded
synthetic-rubber plug into a section of reduced diameter in the piston
guide. The reduction in diameter of the plug increases the travel of the
piston, which is utilized to operate a valve disc, a mechanical linkage,
or an electric switch. The piston moves against the force of a return
spring which causes the return movement of the piston and forces the
thermostatic materials back to their original positions as the element
cools.

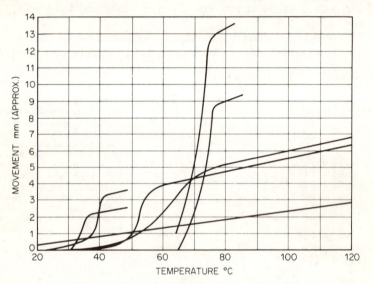

*Fig. 3.7. Typical performance curves for solid/liquid phase wax actuators. (Vernatherm,
American-Standard.)*

   The thermal material starts melting at a given temperature which can
be varied by changing the composition of the wax and the proportion of
copper filler. The latter consists of copper particles dispersed uniformly
in the wax to improve the thermal conductivity. The copper also helps
to vary the amount of expansion with temperature since the greater the
number of copper particles used the less wax is present and, therefore,
the amount of expansion obtained for the same given total volume will
be less. Various compositions are available which have different values
of the critical melting point at which maximum thermal movement
occurs. By controlling the composition, optimum movement can be
obtained over a range as little as 2·8°C, or as much as 56°C. The graph
in Fig. 3.7 shows typical ranges used in commercial production. The
initial gradual slope represents the expansion that takes place whilst the

wax is in its solid state; the sharp rise in movement occurs during the transition from the solid to the liquid state, and the final tail-off shows the expansion that continues after the wax has become fully molten.

By enclosing this thermal movement in a strong outer container, and ensuring that this is able to withstand the internal pressures created by restrained movement of the thermal material, very high pressures can be opposed by the thermal movement and considerable force developed. A large variety of displacement-temperature characteristics are made possible by changing the composition of the wax, the wax-to-copper ratio, and the shape and size of the container and the piston.

Elements of this type find extensive use in conditions where single-point operation is required of a thermal element exposed to varying pressures and vibration. They are especially useful for working in the coolant systems of internal-combustion engines, heat exchangers and domestic and commercial washing machines.

VAPOUR-PRESSURE SYSTEMS

The operating principle of vapour-pressure systems is simple. A limited amount of liquid is enclosed in an expandable capsule such as a bellows, but the liquid does not fill the capsule. On a rise in temperature the liquid will expand but no movement of the capsule will take place as the liquid will merely expand into the space left. However, as the temperature around the capsule passes the boiling point of the liquid, vapour will be driven off its surface. This being unable to escape into the atmosphere will be contained within the capsule. Further rises in temperature will cause increased amounts of vaporization, and a pressure will be built up within the hermetically sealed capsule, causing it to expand (Fig. 3.8).

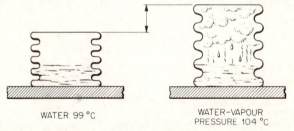

WATER 99 °C

WATER-VAPOUR
PRESSURE 104 °C

Fig. 3.8. Vapour-pressure system. At the left the bellows is at rest with water at 99°C partly filling it. On the right is shown the same system heated to 104°C: trapped vapour pressure has made the bellows expand and move forward.

In principle most of the many liquids that vaporize at a useful working temperature can be used, but in practice the selection is generally

limited to those which are non-toxic, non-inflammable, and capable
of producing a useful vapour pressure over the temperature range for
which the thermostat is being designed. Fig. 3.9.

A system as simple as the one just described would have very limited
application, and an even more limited life. In the first instance all the
useful movement of the capsule would be provided at the boiling point
of the liquid employed. Further rises in temperature would take the
capsule beyond its prescribed movement. Secondly, in the system as

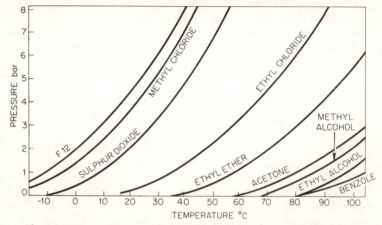

*Fig. 3.9. Approximate pressure characteristics for various liquids used
in vapour-pressure systems.*

described the bellows would always be operating at a length greater
than its free length, and, as pointed out in Chapter 2, this would
seriously affect its life. Most systems, therefore, incorporate an opposing
spring, which overcomes both these limitations (Fig. 3.10).

*Fig. 3.10. Basic elements of a
vapour-pressure system incorporated into a thermostat.*

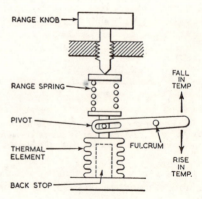

The range spring (Fig. 3.10) compresses the bellows to a length less than its free length, and any thermal movement therefore moves it between this point and its free length, against the opposition of the spring. Limiting stops are generally arranged for both its maximum compression and its maximum extension. The spring also overcomes the drawback of a single operating point, for with a means of setting the opposing spring pressure the temperature at which the liquid will produce sufficient vapour pressure to overcome the spring pressure and move the capsule forward can be determined. Thus we have an immediate advantage over the hydraulic system in that the whole of the movement obtainable from the flexible bellows can be restrained until the operating temperature is reached, and then within a few degrees can be made available for operating a thermostat mechanism.

Two main types of thermal systems are in use utilizing the vapour-pressure system, namely, the self-contained unit and the unit with a remote bulb.

Self-contained system

The self-contained system uses the bellows only, partly filled with liquid, and this becomes the sensing element. Such an arrangement can be found in several types of room thermostat where the bellows and switch mechanisms are all housed within a ventilated compartment designed for wall mounting. Similar arrangements can be found in other applications where it is an advantage to have the sensing device in close proximity to the switch. In these systems the temperature surrounding the bellows is conveyed to the liquid sealed within it. As the temperature rises above the boiling point of the liquid, vapour is driven off the surface creating a pressure within the bellows. The spring opposing the bellows will at first exert sufficient power to hold it back, but further temperature rise will cause an increase in vapour pressure to overcome the spring force and push the bellows forward. At this point the bellows will normally actuate a control on the source of heat causing the temperature rise to halt. Eventually the temperature will commence to fall and, when this change is conveyed to the liquid, the vapour pressure will fall accordingly. With the reduced pressure in the bellows the opposing spring will again exert its authority and be able to compress the bellows back to its back stop.

System with remote sensing element

This self-contained system is somewhat limited in its application and it is often necessary to arrange the sensing of temperature changes to be some distance from the control or thermostat. In these instances it is convenient to use a bellows for the provision of the actual thermal movement, but to have the sensing liquid in a remote bulb or phial, connecting these two with flexible capillary tubing. Thus temperature changes at the bulb create changes in vapour pressure which are then transmitted to the bellows through the capillary.

It has already been explained that it is the temperature change that causes the liquid to evaporate and create vapour pressure. It follows that the surface of the liquid, i.e. the site of the vaporization, must always be in the bulb or at the temperature-sensing part of the system. This aspect needs careful consideration where a separate bulb is used because in some instances the bellows will be colder than the phial and the liquid will condense into the bellows, and in other applications the bulb will always be colder than the bellows, in which case the liquid will be entirely in the bulb. And then there are some applications where the bulb is sometimes hotter than the bellows and sometimes colder.

Low temperature thermostats

Consider first the problems associated with the refrigeration thermostat, where the bulb is always the coldest spot in the thermal system, and where there is the additional problem that when the thermostat is lying in stock the whole system may be subject to temperatures well in excess of the normal working range. Here, as shown in Fig. 3.11, only a small volume of liquid need be used, for it will always be condensed within the bulb which will therefore always enclose the surface of evaporation. The small amount of liquid is also arranged as a means of protecting the system against temperatures much higher than the normal range.

Imagine a refrigeration switch designed to control over a range of $-23°C$ to $+7°C$. A possible choice of liquid that would provide a useful working pressure over this range is Freon 12, with a pressure of 0.27 bar at $-23°C$ and 2.9 bar at $+7°C$. The bellows will be in the thermostat body situated outside the refrigerated zone in which the bulb is placed. Thus in normal operating conditions the bulb, being at a much lower temperature than the bellows, will contain the liquid, and changes of temperature round it will create various pressures which will be conveyed to the bellows and cause this to operate the thermostat and control at a temperature selected by the range mechanism. However, with the refrigerator switched off, or with the thermostat in a store awaiting dispatch, it is quite possible that the whole bellows and thermal system could be subject to a relatively very high ambient temperature of the order of 32°C. At this temperature Freon 12 would be exerting a pressure of 7 bar which might well be in excess of the safe pressure for the bellows of a refrigerator thermostat. It is therefore necessary to arrange that before this internal pressure is reached the whole of the liquid has boiled dry, so that further temperature rises merely cause the vapour to expand in accordance with the laws of gas expansion. As gas expansion is considerably less than saturated vapour-pressure expansion, rises in temperature merely cause small rises in pressure (see Fig. 3.12). This type of charging is known variously as limit charging, fade-out charging, or gas charging.

In some thermostats, especially those used in domestic refrigeration, this type of charging is carried a degree further so that the control range is entirely within the gas-pressure range. In these thermal systems vapour is introduced at a known pressure and temperature and the

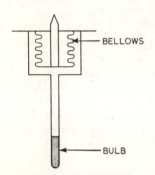

Fig. 3.11. *Vapour-pressure or gas-charged system for low temperature applications.*

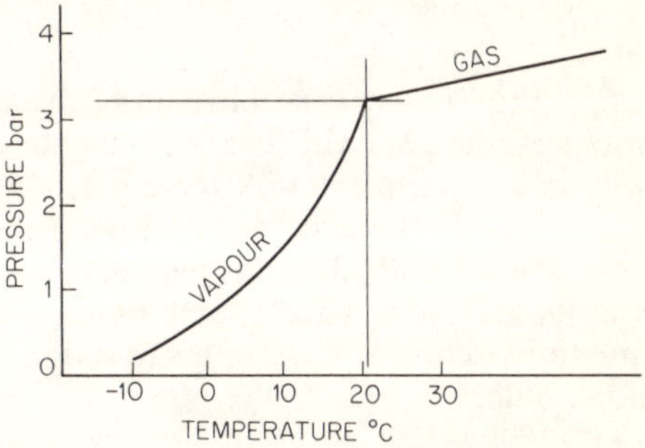

Fig. 3.12. *Graph to show the effect of limit charging where, at a predetermined temperature, all the liquid is boiled off after which further rise in temperature will superheat the saturated vapour causing further pressure increases to follow gas expansion laws.*

thermal range is chosen so that this is above the point where saturated vapour is present: all temperature selection takes place at points where the vapour has become dry and behaves as a superheated vapour or gas. These systems are very sensitive because there is very little thermal mass to be subjected to the change in temperature. With the absence of large amounts of thermal liquid the sensing bulb can become part of the capillary and this too contributes to the sensitivity of the instrument because the thermal mass normally associated with the large sensing bulb of the vapour-pressure system is no longer present.

High-temperature thermostats

The second class of vapour-pressure system concerns applications where the bellows is always colder than the sensing bulb—systems used in central heating, ovens, and air heaters, where the thermostat has a

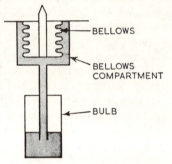

Fig. 3.13. High temperature system where it is assumed that the bellows is always the coldest point in the system and therefore liquid will always be condensed back to the bellows. Sufficient liquid is therefore necessary to fill the bellows, capillary and part of the bulb.

temperature range well above ambient, are typical examples. In these instances, the bulb will always be the hottest part of the system and the bellows compartment will, therefore, always be filled with condensed liquid. As the bellows moves forward under pressure, more liquid will leave the bulb, and it is necessary to ensure that when the bellows is at its maximum stroke some liquid still remains in the bulb and is able to present a surface of evaporation to the changing temperatures in the bulb.

Systems where bellows is colder or hotter than the bulb

The third type of vapour-pressure system for which special precautions are necessary is where the bellows, or other type of capsule, can be either colder or hotter than the sensing phial. If, for instance, this condition occurred in the refrigeration system previously described, some difficulties could be encountered. As previously stated, the thermal charge will always condense at the coldest spot of the thermal system. In the case of the refrigeration thermostat the amount of liquid is kept purposely small, and therefore if the bellows becomes the coldest spot, all the liquid will condense in it and there will be none left in the bulb. The bellows end of the thermal system would then become the temperature-sensing point, and any change of temperature at the sensing phial would have no effect whatever. As it happens, in the refrigeration thermostat we can tolerate such a possibility, for in normal working conditions the bellows will be sited in the thermostat switch assembly which will always be warmer than the sensing phial

sited either in the refrigerated area or in close proximity to the evaporator (in both cases the sensing phial will be the coldest point of the
system).

In a thermostat controlling at ambient, or near ambient temperatures,
however, such a thermal system would present some difficulties. A typical
example would be a thermostat holding a process liquid bath at 18°C.
In summer the thermostat housing the bellows would be in ambients
of 21°C and even higher. In this condition the liquid would be condensed in the phial, which would be the coldest spot; whilst in the
winter ambients of the order of −1°C could be experienced, in which
case, the bellows being colder than the process, the liquid would condense and remain in the bellows.

To meet this possibility, enough liquid must be put into the system to
ensure that some is always left in the phial. Such a system is shown in
Fig. 3.14. Here it will be seen that with the bellows chamber the coldest

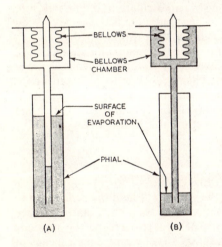

BELLOWS

BELLOWS
CHAMBER

SURFACE
OF
EVAPORATION

PHIAL

(A)   (B)

*Fig. 3.14. Cross ambient systems. For
applications where either the bellows
or the sensing element can be the coldest spot sufficient liquid has to be available so that: (A) when the phial is coldest all the liquid in the system can be
contained in the phial, or (B) when the
bellows is the coldest point in the system it can draw liquid out of the phial
and still leave some liquid in the phial
to sense changing temperatures.*

spot in the system, liquid is drawn up into it (B). Even so, some liquid
remains in the phial, and temperature sensing is retained at the surface
of evaporation in the phial. Changing temperatures cause changes of
vapour pressure in the remaining space in the phial, and these pressure
changes are transmitted to the bellows hydraulically via the liquid in
the capillary and in the bellows chamber. Should the temperature round
the bellows rise above that surrounding the phial the liquid will condense back into the phial. It is essential to ensure that the surface of
evaporation is retained in the phial, and so it is necessary to ensure that
the phial is of sufficient volume to hold all the liquid and still provide a
space above the surface, as shown in Fig. 3.14. When calculating the

amount of liquid required it must be remembered that changes in temperature will cause the liquid to expand or contract. Therefore the volume of the phial or bulb must be such that when the liquid in the phial is at an elevated temperature the phial can accommodate all the liquid at its greatest volume and still provide some area above the surface of evaporation.

In this type of vapour-pressure system the point at which the temperatures of the bulb and bellows are equal is the most critical and some temporary loss of control may be encountered. The change of conditions causing the liquid to condense back from the bellows to the phial or vice versa will cause some fluctuation in the vapour pressure, which will result in the thermostat operating erratically. The period during which this occurs depends largely on the speed at which the changed temperature condition takes place and the length of the capillary. A quick change in temperature in a system with a metre or so of capillary will probably cause one or two oscillations of pressure, after which stability will again be established; but a gradual change of a few degrees of temperature cross-over in a system with a lengthy capillary could result in loss of control for several minutes. Because of these limitations, most manufacturers and users of controls avoid as far as possible vapour-pressure systems when such a cross-over can occur. By careful siting of the thermostat it can sometimes be ensured that the bellows is always the hottest point of the system, even if it means resorting to some method of conducting heat to the bellows.

Vacuum charge

In any vapour-pressure system that employs an opposing spring to obtain a useful operating range, the spring will hold the bellows on to its back stop and prevent it from moving until the liquid has been heated to develop a vapour pressure sufficient to overcome the opposing spring pressure. This has the advantage that by providing a means of presetting the opposing spring pressure a range of operating points can be obtained from one single thermal system. Unfortunately, should the thermal system be damaged, and the thermal charge escape, there is the disadvantage that the spring will hold the bellows back indefinitely until the faulty system is detected and replaced.

In some instances the faulty condition could be dangerous. In a car-engine cooling-system thermostat, for instance, where on a rise in temperature the bellows has to move forward and allow the coolant to circulate through the radiator, loss of thermal charge in a spring-opposed bellows would mean that the bellows would never move forward and the car engine would become seriously over-heated because of the failure of the thermostat to allow the cooling water to circulate. A similar fault occurring in an oil-burner control would mean that the burner would be left on indefinitely by the failure of the bellows to move forward on a rise in temperature and cut off the oil supply to the burner; this condition too could have dangerous consequences. In such cases, therefore, provision must be made so that in the event of a loss of thermal charge the bellows automatically moves forward or, as it is frequently termed, 'fails safe'.

In a single-point temperature controller such as a car-coolant thermostat the solution is comparatively simple (Fig. 3.15). The opposing spring is dispensed with and the bellows is virtually unrestricted. A thermal liquid is chosen which boils at a temperature equal to the temperature at which movement is required to begin. The bellows is partly filled with the liquid and it is arranged that the space left is evacuated of air. In this manner at temperatures below the boiling point of the liquid the bellows is collapsed on to its back stop by atmospheric pressure. On a rise in temperature above the boiling point of the liquid, vapour pressure is created and eventually equals the opposing atmospheric pressure. At this point the spring rate of the bellows is able to

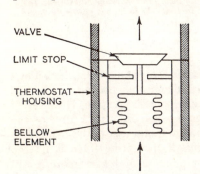

Fig. 3.15. Vapour-pressure system using vacuum charge for fail-safe conditions. Diagram shows a typical example of this type of control in the cooling system of an automobile engine.

assert itself and allow the bellows to move forward to its free length. It is generally arranged that at this point the bellows comes in contact with a forward limit stop. Further rises in temperature will then merely increase the internal vapour pressure but no more movement takes place. The advantage of this particular arrangement is that if the bellows becomes punctured and the thermal charge escapes, air will enter the bellows and there will be a consequent loss of vacuum. The pressure internally and externally then being balanced, the bellows will move forward under its own spring rate to its free length position, and, in doing this, will be caused to do the same work as it would normally be forced to do in the event of a temperature rise. It would, of course, stay in that position until rectified, or replaced; but in some instances, for example that of the car engine cooling system thermostat, this is much to be preferred to a bellows that fails and remains in its compressed condition.

It is possible, to a limited degree, to combine spring-controlled range selection with this failure-to-free-length feature of the bellows. Fig. 3.16 shows the thermal system of a Teddington oil controller which incorporates these two features. Here it will be seen that the bellows normally pushes a lever which in turn operates a valve or switch. The bellows system is charged in a similar manner to the first example in that a liquid is chosen where boiling commences at the temperature at which operating conditions will apply. The space left in the thermal system is evacuated so that at ambient temperature the bellows will be opposed

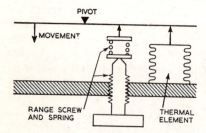

Fig. 3.16. *Vapour-pressure system using vacuum-charged thermal element offering fail-safe feature. Diagram shows method of incorporating range adjustment. In this instance the bellows is positively linked to the lever.*

and compressed by atmospheric pressure. Normally, then, the bellows will only move when the liquid has boiled and when the vapour pressure has developed a pressure equal to atmospheric pressure. However, it will be noted that a spring is introduced which exercises a force on the lever in the same direction as the bellows, and assists the bellows to overcome atmospheric pressure. Thus if the spring force is nil the bellows will need to await a vapour pressure of 1 bar before it will

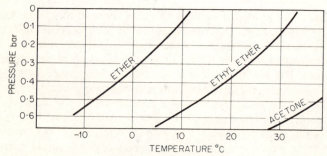

Fig. 3.17. *Pressure curves for liquids that can be used in a vacuum-charged system.*

move forward, but if the spring force equals some 0·7 bar then the bellows will only need a smaller vapour pressure plus its own spring rate to move forward. A glance at the graph (Fig. 3.17) shows that there are several liquids which provide useful vapour pressure ranges for such applications, even though the total range from any one of these is somewhat limited. Thus, in such a system range selection can be provided and the 'fail-safe' feature (in that the bellows moves forward if it looses charge) is retained.

Effect of variations in atmospheric pressure and phial level

In vapour-pressure systems the pressure developed by the thermal charge has to overcome two opposing factors: (1) the pressure of the range-adjusting spring; (2) atmospheric pressure.

In most applications a liquid is chosen to provide the thermal charge which produces pressures well in excess of atmospheric pressure. Thus any day-to-day change of atmospheric pressure represents such a small proportion of the total pressures involved that the error introduced by these barometric changes is comparatively small. The only systems likely to suffer from the effect of barometric pressure changes are those vacuum-charge systems, described above, where a fundamental feature of the operation is that at normal ambient temperatures the bellows is under vacuum and is pressed backwards on to its backstop by atmospheric pressure. Obviously, on these systems a fall in barometric pressure is likely to allow the bellows to move more easily off its backstop, and therefore will materially affect the operating point of the thermal system as a whole.

The same vacuum-charged thermal systems are also likely to be affected by the head pressure of the thermal liquid on the bellows. When the sensing phial is level with the bellows there obviously will be no head of liquid pressing on the bellows, but if the sensing phial is lifted above the bellows the head pressure caused by the liquid could materially affect the performance of the thermal system as a whole.

In commercial instruments for general duties these deviations are normally tolerated, but for scientific or laboratory use instruments of this type are generally provided with a correction factor, so that their performance can be checked against the barometric pressure of the day, or can be re-calibrated to compensate for differences of head pressure caused by lifting the sensing phial above the bellows.

Advantages and limitations of vapour-pressure systems

The vapour-pressure system offers a means of temperature sensing that can be fairly easily produced in large quantities, and with economics that allow its use for low-priced controls. It also has the advantage that with a properly constructed remote bulb and capillary system only the sensing bulb reacts to temperature change and, therefore, head sensitivity is eliminated.

Difficulties associated with the need for ensuring that the evaporating surface is always in the sensing bulb have already been discussed, as have those associated with cross-over conditions. There is also the disadvantage that a vapour-pressure system, following the vapour-pressure-temperature curve, will not produce a linear scale. As most range knobs rotate a screw thread of constant pitch, the spring that is being compressed by this rotation will be loaded at a constant rate of increase. The bellows being opposed, however, will only react to the vapour-pressure curve, and therefore the angular rotation of the knob could produce fewer degrees temperature change per degree angular rotation at the bottom of the range than at the top.

The deviation from the linear will depend on the vapour-pressure curve for the liquid used and the section of the curve over which the temperature range is selected. This limitation is generally accepted by the thermostat designer who arranges for the range knob, or scale marking, to vary in relation to the pressure curve.

Non-linearity of temperature selection can also be overcome by the use of a cam instead of a screw thread, the cam being shaped to cancel out the non-linearity of the temperature curve.

ADSORBER CHARGE

This type of system consists of a sensing phial filled with a solid adsorber medium which generally takes the form of activated charcoal, the whole system subsequently being charged with an inert gas such as carbon dioxide or nitrogen. The principle used in this system is that at low temperatures the adsorber takes up the gas and in so doing reduces the pressure in the system. On a rise in temperature the gas is released from the adsorber and creates in the thermal system an increase in pressure, which depends on the temperature at which the adsorber in the sensing phial is held.

Various temperature-pressure ranges can be produced, depending on: (1) the total volume of the system; (2) the amount of adsorber in the sensing phial; and (3) the temperature and pressure of the gas at the time of the initial charging.

Manipulation of these factors can produce a variety of temperature and pressure curves.

4

THERMOCOUPLE AND ELECTRICAL RESISTANCE THERMAL ELEMENTS

THERMOCOUPLE ELEMENTS

IT has long been known that a loop consisting of two wires of different metals welded together at their ends will, on being heated at one weld or junction, produce an electromotive force (e.m.f.) and cause a current to flow round the loop. It is also known that the amount of current flowing bears a consistent relation to the temperature difference between the two junctions. This basic principle is incorporated in the thermo-couple and its application in temperature measurement and control represents an important section of the range of devices available for heat control work.

Fig. 4.1. Operating principle of thermocouple circuit. Heat applied to one end of the thermocouple, causing a temperature difference between the hot and cold junction, sets up an e.m.f. which can be indicated on a galvanometer in the circuit.

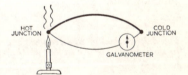

Basic circuit

The basic circuit of a thermocouple is shown in Fig. 4.1. Two wires of dissimilar metals are joined together at their ends and are for the rest of their length held apart. One joint is subjected to the higher temperatures required to be measured, or controlled, and is known as the hot junction. The other joint is kept relatively cool and is termed the cold junction. In this basic circuit current will flow at a rate roughly proportional to the temperature difference between the two junctions and, with suitable precautions, the temperature difference can be deduced from measurements of the current. Obviously, if the temperature of the hot junction is to be accurately measured in this way it is necessary to arrange for the cold junction to be kept at a constant temperature or for some compensation to be provided to offset any temperature change at the cold junction.

Compensating for changes in cold-junction temperature

In laboratory practice this is usually achieved by immersing the cold junction in melting ice or boiling water; in most commercial applications the cold-junction temperature is allowed to vary with that of its surroundings, and the amount of the change is either allowed for by direct measurement or compensated for by some other thermally sensitive device. In the former case, the e.m.f. between the hot and cold junctions is measured, and the temperature of the cold junction is then taken by thermometer. The e.m.f. figure proportional to the cold-junction temperature, obtained by reference to the calibration curve for the particular type of thermocouple, is added to the first e.m.f. reading, and the temperature of the hot junction can then be calculated again using the calibration curve.

For a continuous reading or controlling instrument provision is made for an automatic compensating mechanism. This continually senses the cold-junction temperature and is able to adjust the resistance in the circuit or to bias a millivoltmeter reading, so balancing out changes in e.m.f. due to changes in the temperature at the cold junction. The recorder or controller will then read the true temperature at the hot junction.

The e.m.f. flowing through the thermocouple circuit can be measured by a galvanometer type of instrument placed in the loop at the cold-junction end.

Fig. 4.2. Method of transferring thermocouple reading to an indicator or controller which becomes the cold junction.

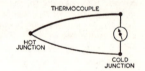

Thermocouple circuits

In the simplest circuit the thermocouple leads run from the hot junction to the indicator, or controller, which then becomes the actual cold junction (Fig. 4.2). This arrangement is frequently used in portable instruments or other applications where the indicator or controller is conveniently close to the hot junction. For other applications where the point of measurement is remote from the hot junction other arrangements can be made to avoid the necessity for running the thermocouple circuit all the way back to the measuring device. In such arrangements it is essential to ensure that the cold junction is either maintained at a constant temperature or is compensated for changes, and that any leads extending the circuit from the cold junction to the indicator do not affect the flow of current in the thermocouple circuit or bias this in any unpredictable manner.

Fig. 4.3(a) shows the simplest method where the cold junction is maintained at the end of the thermocouple, and extension leads, generally of copper, are then carried to the measuring instrument. This type is suitable where it is possible to maintain control over the cold junction temperature at a place near the hot junction. Where this is not possible the use of compensating leads is employed, these leads being of the same or similar material to the thermocouple wires themselves. In this manner the cold junction can be carried back to the instrument and control or measurement of the cold junction temperature is then incorporated in the measuring instrument (Fig. 4.3(b)). In fact, a large number of applications use a combination of these two methods

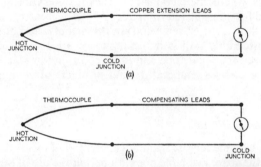

Fig. 4.3. (a) *Arrangement where the cold junction is maintained at the thermocouple, but readings are conveyed to a remote indicator or controller via copper extension leads.* (b) *Use of compensating leads allows for remote indication or control from the thermocouple and transfers the cold junction back to the indicator or controller.*

where expense (in the case of rare metals) prohibits the use of compensating leads all the way back to the instrument. In these instances compensating leads are used to extend the thermocouple circuit back to a cold junction remote from the hot junction, and then copper leads are used from this point back to the measuring instrument.

Construction

The materials used in the manufacture of thermocouples fall into two groups, namely rare metals and base metals. In the rare metal group the positive wire is of platinum-rhodium alloy and the negative wire of platinum. In the base-metal group there is a variety of metals including copper-constantan, iron-constantan, and chromel-alumel. These metals are chosen as in general they meet the essential requirements of thermocouple metal: (1) an approximately linear e.m.f.-temperature relationship; (2) a constant e.m.f.-temperature relationship throughout the useful life of the thermocouple; and (3) as large an e.m.f. as possible to facilitate accuracy of measurement and control.

The hot junction of the thermocouple is generally made by twisting the two dissimilar metals together and then welding. In making this joint it is essential that there is negligible resistance through it when welded and that a firm bond is made between the two metals so that subsequent fracture or increase of resistance is prevented. In another form the thermocouple takes the shape of a tube, a typical example being the type used as a thermoelectric flame-failure device. These thermocouples are either of the iron-constantan type or of other special alloys, where one metal is used for the outer tube and the other metal is used for an insulated rod passing down its centre, the two being welded together at the extreme end. In the manufacture of this type of element considerable skill is necessary to ensure that in use the connexion between the outer tube and inner rod is not gradually corroded by the action of high temperature and products of combustion from the flame.

Thermocouple applications

The thermocouple represents an ideal method of measuring temperatures where the space available does not permit the use of other types of thermally sensitive device. In addition its low thermal mass ensures rapid sensing of temperature changes and its construction allows temperatures to be measured up to 1600°C or more, according to the materials used. In the case of iron-constantan the range is 150°C–750°C, and in the case of copper-constantan 180°C–320°C.

With special metals, such as tungsten-molybdenum, up to 2600°C can be measured, although this last material is only used on rare occasions.

The thermocouple can be used for the measurement of temperature in several ways. The simplest method is to place the bare thermocouple in contact with the surface or liquid whose temperature is being measured. Such a method is frequently used in laboratory checks on surface temperatures, on domestic appliances, in performance checks on other thermally controlled liquid baths, and for most applications where the thermocouple is not subjected to corrosive action during its application. Where the conditions are such that the thermocouple would deteriorate rapidly if not protected, a protective immersion tube is generally built round the thermocouple. In these applications the thermocouple wires themselves are first insulated one from the other with a material of sufficient strength and thermal property to ensure that good insulation is maintained under working conditions. It is then placed in the immersion tube where steps are taken to ensure that the thermocouple is in good firm contact with the tube so that temperature changes of the medium being measured can be transferred to the thermocouple with as short a time lag as possible.

Great care should be taken on all the wiring between the thermocouple and the measuring instrument. This is necessary to ensure that no false e.m.f. arises in the circuit. The insulation of the thermocouple, the compensating leads and the extension cables must be sufficiently sound to withstand normal usage without danger of a leakage current through it.

Compensating and extension leads should be made in single lengths with no joins other than those necessary to complete the circuit.

Various methods are adopted whereby the e.m.f. flowing through a thermocouple in relation to temperature at the hot junction can be measured and converted into temperature indication, recording, or control. These methods generally fall under two main headings, i.e. direct-deflexion and potentiometer instruments.

Direct-deflexion instruments

In this method, a moving-coil galvanometer, which is inserted in the thermocouple circuit and acts as a direct-deflexion millivoltmeter, is used. These instruments are calibrated by reference to standard tables giving the e.m.f./temperature relationship of the thermocouple being used. Various types of millivoltmeters are used for the direct-deflexion instrument to meet the particular requirements of the external circuit, and each manufacturer makes his particular recommendations for the type of instrument best suited to the thermocouple application and circuit (Fig. 4.4). In most direct-deflexion instrument systems the cold junction is carried back to the meter and this becomes the cold junction itself, so that compensation for variations in cold junction temperature can be incorporated into the meter. The galvanometer or millivoltmeter is of the moving-coil type, with the coil delicately balanced and normally retained in position by means of hair-springs. It is therefore possible to incorporate cold-junction compensation at this point by making part of the hair-spring system of bi-metal. Any change in temperature within the latter will cause the bi-metal to warp and bias the spring balancing of the moving coil. In this way any false e.m.f. introduced by changes in temperature of the cold junction will be automatically cancelled out, and the indication will be the true hot-junction temperature (Fig. 4.4).

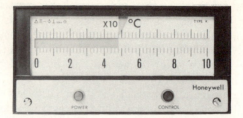

Fig. 4.4. Typical potentiometer used for temperature indication and Control. (Honeywell Ltd.)

Potentiometer instruments

This type of instrument uses a potentiometer circuit and can be of the true-null or the balanced type. In the true-null type the e.m.f. generated by the thermocouple disturbs the balanced circuit, the balance being restored by manual operation—with reference to the centre zero galvanometer. The amount of adjustment required to restore the balance is then noted and the figure converted to temperature. Other types of null-balance instruments can be read off in terms of temperature without the necessity of converting the e.m.f. figure. The true-null instrument under normal conditions is independent of the resistance of the external circuit.

Fig. 4.5. Potentiometric indicator-recorder. (Honeywell Ltd.)

The second type of potentiometric instrument used in thermocouple work is the self-balancing variety. In this type states of unbalance created by changes in the e.m.f. across the thermocouple are automatically balanced out as they occur. In the balanced condition the e.m.f. across the thermocouple circuit is 'balanced' with a current from a standard cell or a stabilized voltage supply so that the galvanometer reading is zero. A change in e.m.f. across the thermocouple circuit, caused by a change in temperature, upsets this balance. A mechanism is set in motion to balance this e.m.f. against the voltage drop across a slidewire. The amount of movement of the adjusting mechanism is indicated by a pointer which is marked out in temperature readings.

Cold-junction compensation is normally incorporated in the potentiometer circuit and consists of a resistance, or a series of resistances, having known temperature-resistance coefficients. Thus, for a change in temperature around these resistances the potentiometer circuit is biased to provide the correct cold-junction compensation.

Recording temperature

In their simplest form thermocouple instruments, both of the direct-deflexion type and the potentiometer type, are used for indication only, but it is not difficult to convert them into recording instruments. In the majority of cases the recording is done on drums or continuous rollers. In the direct-deflexion type instruments where the pointer is part of the galvanometer system, the lack of power available precludes the use of continuous pen and ink recording. This type of instrument therefore uses a chopper bar and an inked thread suspended a short distance above a roll type chart. The indicator or pointer assumes its position in relation to temperature, and is at intervals pressed down by the chopper bar, in turn pressing the inked thread against the paper. By substituting a multicolour typewriter styled ribbon and adding a motor driven selector with a multipoint recorder it is possible to point out a number of recorders identified by colour. Owing to the frequency with which the chopper bar operates a nearly continuous line is recorded on the chart, the intervals when the chopper bar is raised still permitting the pointer to assume its correct position relating to the temperature as indicated by the thermocouple.

In the potentiometer self-balancing type of instrument with amplification, considerable power is available for the recording and controlling mechanisms. Various types of control can be superimposed on these instruments, ranging from on-off switching with controllable differential, floating control giving two or three speeds, and proportional control with infinitely-variable throttling curves. These various types of control can transmit signals to valves and other power-operated devices, and thus provide a complete circuit of indication, record, and control. A motor drive selector switch and indexing mechanism will enable a number of variables to be recoded on the one chart identified by the appropriate colour and/or symbol.

ELECTRICAL RESISTANCE THERMAL ELEMENTS

This type of sensing element relies on changes in the electrical resistance of a conductor (usually a metal wire) with changes in the temperature surrounding it. Various metals are used for the wire including platinum, nickel and copper—depending on the temperature over which the element is required to operate.

Resistance bulbs are used extensively for low-temperature/low-span applications and offer greater accuracy than thermocouples over a working range of $-200°C$ to $+500°C$. They are frequently used in electronic systems for air conditioning and are equally suitable for temperature indication and control where a number of remote points of

Fig. 4.6. Indicator used with resistance-element circuitry. (Honeywell Controls Ltd.)

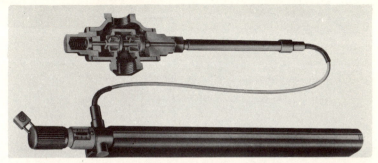

temperature measurement are required to be displayed on one central indicating and controlling station.

The Wheatstone bridge type of measuring circuit is used where the resistance element takes the place of one of the resistors in one arm of the bridge. A change of resistance at the sensor, due to a change of temperature, produces an out of balance current in the bridge. The null-balance potentiometer method is used to detect and restore the electrical balance, the necessary adjustment being made either manually or automatically. A recorder for use with resistance thermal elements is shown in Fig. 4.6.

As with thermocouple instruments, resistance elements can be made to work in conjunction with indicators, indicator-recorders, or indicator-recorder-controllers. Suitable amplifiers are supplied by each manufacturer so that any type of control can be superimposed on the circuit, including on-off or two-position control where both the on and off points are positioned at predetermined points along the scale. Floating control can be provided where the signals for varying rates of response are set at predetermined points along the range; on the other hand, proportional control can be provided by which over a given section of the temperature scale a proportioning potentiometer is brought into play which, in turn, transmits modulating or proportioning action to valves and damper units.

TRANSMISSION, RANGING COMPONENT, ELECTRIC SWITCHES, AND CONTACTS

TRANSMISSION AND RANGING COMPONENT

THE movement generated by the thermal element in a thermostat has to be transmitted to a valve or switch mechanism: it is very rarely possible to connect the thermal element directly to either a switch or valve. One or more intermediate links are generally required to convey the thermal movement to the controlling member in an acceptable manner. This linkage, in its various forms, is known as the *transmission component*.

Most types of thermal element provide a useful movement in relation to temperature over a comparatively wide temperature band. Thus in most thermostatic devices a means of range and set-point adjustment is required so that the movement provided by the thermal element can be made use of only at that temperature at which the valve or switch is to operate. For this purpose the transmission component generally incorporates two other features: (1) some means of absorbing or restraining the thermal movement until a predetermined set-point is reached, and (2) a method by which the set-point can be predetermined and adjusted over a temperature range. The transmission component may also include a means of multiplying the thermal movement, or converting a proportional movement into a snap-action movement.

Absorbing or restraining thermal movement

The first function of the transmission link between the thermal element and the control valve or switch is to absorb the thermal movement or restrain it until the set-point has been reached. Whether the thermal movement is absorbed or restrained depends on the type of thermal element employed.

Where the movement provided by the thermal element is hydraulic expansion, which it would obviously be impossible to restrain, some means has to be incorporated of absorbing the unwanted movement provided when the element is above or below the control point. A method of doing this is shown in Fig. 5.1. Range adjustment is carried out by rotating screw *A* in the head of the sensing phial. This moves push rod *B* in the bellows, causing liquid to be displaced along tube *D* and transferred to the thermostatic valve *C*, which is then pre-set in the desired position in relation to the valve seat. When this valve is forced on to its seat by thermal expansion, any further expansion is absorbed in the sensing phial by the spring between *A* and *B*.

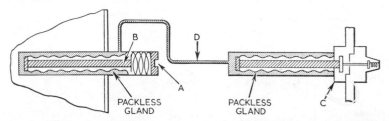

Fig. 5.1. *Hydraulic transmission unit showing how thermal expansion of liquid is transferred from a sensing phial to a valve, causing the valve to move to the closed position.*

In vapour-pressure systems, on the other hand, the transmission link can be arranged so that it restrains thermal movement until the correct temperature-pressure relationship for the set-point has been reached, when movement is passed to the control component. This is generally achieved by combining the restraining action and temperature setting-point mechanism in the form of a spring (see Fig. 5.2). By pre-setting

Fig. 5.2. *Vapour-pressure system. In this type of thermal unit the bellows under the influence of rising vapour pressure is allowed to move forward until it engages the forward stop. After that, any excess temperature merely causes increasing vapour pressure to build up in the bellows without forward movement. The limiting factor is determined by the maximum pressure the bellows can withstand without permanent deformation.*

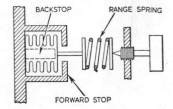

the spring pressure opposing the bellows or diaphragm in the vapour-pressure system, thermal movement can be prevented until the vapour-pressure has risen to balance out the spring pressure, after which further rises in temperature will overcome the opposing spring pressure and allow the thermal element to convey movement to the transmission link and the control component.

Typical examples of these two thermal systems applied to direct-acting valves are shown in Fig. 5.3 and Fig. 5.4. In the former, a vapour-pressure system is arranged to operate a valve: on a rise in temperature the bellows moves forward and closes the valve against a restraining spring. The compression of the spring is initially adjusted so that the vapour pressure in the thermal system will be insufficient to move the bellows forward and close off the valve until the operating temperature is reached. In Fig. 5.4 a similar valve is controlled by a hydraulic thermal system, and no attempt is made to restrain the movement of the thermal unit. Instead, the valve is attached to the

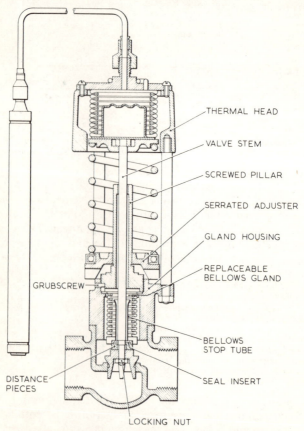

THERMAL HEAD

VALVE STEM

SCREWED PILLAR

SERRATED ADJUSTER

GLAND HOUSING

REPLACEABLE BELLOWS GLAND

GRUBSCREW

BELLOWS STOP TUBE

DISTANCE PIECES

SEAL INSERT

LOCKING NUT

Fig. 5.3. Typical self-contained thermostatic valve operated by a vapour-pressure system. (Teddington Autocontrols Ltd.)

thermal unit and the position of the valve in relation to its seat is initially adjusted by the ranging mechanism so that the thermal movement is just enough to close the valve at the desired temperature. Since the thermal sensing liquid is incompressible some safety arrangement has to be provided so that once the valve is shut any movement

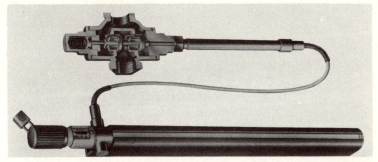

Fig. 5.4. Direct-acting thermostatic valve using a hydraulic expansion system. (Sarco-Spirax.)

due to a further rise in temperature can be absorbed. Temperature over-run protection in Fig. 5.1 is provided by means of a heavy spring in the setting head. This arrangement can, however, lead to a change in the set point of the instrument if the spring takes a permanent set over a period of time. Over-run protection in the instrument in Fig. 5.4 avoids this problem. A sealed bellows filled with an inert gas under pressure is built into the temperature-sensing bulb. Excess oil pressure generated by temperature over-run is then taken up by compression of the bellows and the thermostat can cope with temperature over-run of up to 38°C above the set point without damage.

ELECTRIC SWITCHES

The same basic problems, of either absorbing thermal movement or restraining it, are present when electric switching is used in thermostats. Apart from simple thermostats calling for single-point operation, the majority require a temperature adjustment over a specified temperature range so that, as with the thermostatic valve, the transmission component must do at least two things: (1) restrain the thermal movement or absorb it until the temperature set-point has been reached, and (2) provide an adjustment so that this set-point can be chosen over the operating temperature band.

Figs. 5.5 and 5.6 show two simple ways of achieving these two functions.

Fig. 5.5. Bi-metal switch with sprung fixed contact to absorb excess thermal movement.

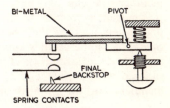

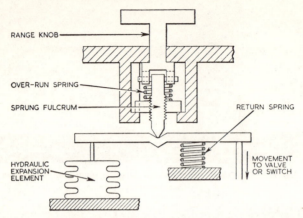

RANGE KNOB

OVER-RUN SPRING

SPRUNG FULCRUM

RETURN SPRING

HYDRAULIC
EXPANSION
ELEMENT

MOVEMENT
TO VALVE
OR SWITCH

*Fig. 5.6. Hydraulic thermal unit attached to an electric switch with
a sprung fulcrum to absorb excess thermal movement.*

Snap-action switches

In addition to these two basic functions the transmission link in an electric thermostat sometimes has to perform others, including the conversion of the thermal movement from a modulating action to a snap action. A case where this is particularly necessary is that of a thermostat incorporating a switch mechanism capable of switching loads of up to 15 A at mains voltage. It would be difficult to make the switch operate satisfactorily under the direct, relatively small and slow movement obtained from the thermal element. The difficulty can be overcome either by considerably magnifying the thermal movement so that the contacts open comparatively quickly, or by converting the proportional thermal movement into a snap action. Both types of mechanism are employed, and Figs. 5.7 and 5.8 show examples.

Fig. 5.7 shows a typical application where a vapour-pressure thermal system is linked to a snap-action switch. It will be seen that the bellows movement is opposed by a spring and a range knob is provided which can vary the force of the opposing spring. These three components operate on a common centre line, and at one stage are linked to a cross member one end of which actuates the moving contact. The range mechanism is coupled to the cross member, but a small amount of play is left between the two, the arrangement being such that as the operating temperature is approached the arm is free to move upwards faster than the normal rate of thermal movement. This allows the magnet situated at the other end of the cross member to take over as the member is drawn upwards, snapping the contacts together. On a fall

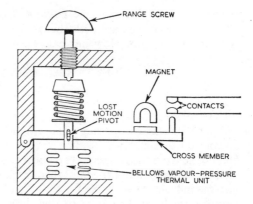

Fig. 5.7. Diagrammatic view of a typical vapour-pressure thermal system incorporating a magnetic snap-action switch.

in temperature the thermal mechanism will attempt to take the moving contact away from the fixed one. Momentarily the force of the magnet will be sufficient to restrain this action and will hold the two contacts in the closed position. Eventually the opposing force of the thermal system will be sufficient to break the force of the magnet and the moving contact will thus snap away from its closed position.

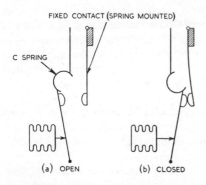

Fig. 5.8. Diagrammatic representation of snap-action switches using a 'C' spring to obtain the snap movement. Note spring action of fixed contact.

Another means of obtaining snap-action is the use of the 'C' spring, generally as shown in Fig. 5.8. Here the over-top-dead-centre principle is employed so that the moving contact is forced into one of two extreme positions, either fully closed and in hard contact with the fixed contact, or fully open against its back stop. In these devices the term 'fixed contact' is sometimes misleading for, in some instances, this is also sprung so that in the closed position (see Fig. 5.8(b)) the moving contact displaces the 'fixed contact'. This enables the fixed contact to move and follow the moving contact in the initial stages of its attempt to open, ensuring that by the time the moving contact is ready to snap fully open it is still in contact with the fixed member. This overcomes the problem of an initial slow movement associated with such devices.

ELECTRICAL CONTACTS

The design of contacts for current-controlling switches used in thermostats presents many difficulties both as regards suitable contact material and also mechanical details. The conditions of switch closing and opening in thermal switches are severely restricted by the movement obtainable from the prime mover. Most thermal movement is linear in relation to temperature rise, and therefore on slow changes of temperature the movement available for transferring to contact movement can be quite small, especially when fine differentials are required. If the contacts are operated directly by this small thermal movement ('creep action') the main problem is to avoid arcing or welding as the contacts slowly approach or recede from each other. If, on the other hand, the slow thermal movement is converted into a snap-action, the total forces available are relatively small, so that contact design must include precautions against low pressure and contact bounce.

Other factors to be considered in the design of contacts for thermal switching include: (1) Resistance to electrical and mechanical wear. This is especially important on thermostats having a high frequency of operation. (2) Resistance against tarnishing of surfaces. This should be borne in mind when considering thermostats for industrial applications where the switch operates in a highly tarnishing atmosphere, and where electrical resistance would be rapidly built up if the contact surface was not resistant to this.

It is generally necessary to compromise between these two main factors and others, including cost, electrical conductivity, and thermal capacity. Thus there is no one metal that can be offered for all applications.

To a certain extent the electrical requirements of the contacts follow a set pattern and Fig. 5.9 shows one manufacturer's recommendations for contact size in relation to contact pressure and load.

Surface tarnish

The prevention of surface tarnish is particularly important in small thermal switches for relay circuits and servo-mechanisms where the actual electrical loads are light and contact pressures small. In these conditions tarnishing can cause intermittent operation of the switch with resultant damage to the main contacts which the relay is control-

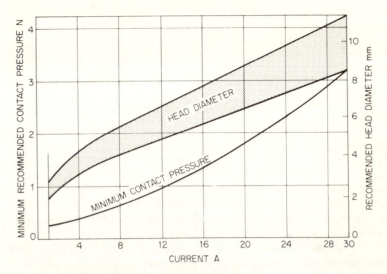

Fig. 5.9. Graph showing contact pressure and head diameter in relation to electrical load.

ling. For heavy-duty switching this problem is not so important, but nevertheless can cause some overheating of the contacts.

Resistance to wear

This is an especially important feature in some thermostat designs where the frequency of cycle is high.

Metals used

Silver is the most widely used metal for contact work, particularly, in its pure form, for heavy-duty thermostat contacts. Where conditions of high oxidation or corrosion (notably in sulphurous atmospheres) prevail an alloy of noble metal may be necessary. More general problems are wear, sticking, and erosion, which are best dealt with by specific alloys or powder-metallurgy combinations. These include silver-cadmium and silver-cadmium oxide, which increase resistance to sticking. Palladium added to silver gives greater resistance to tarnishing and wear, but this lowers conductivity and may reduce the current rating of the thermostat.

6

THERMOSTATIC CONTROL
DEFINITIONS

ANY heating or cooling process can be automatically controlled by the use of thermostats and associated equipment, so long as the medium which is to be controlled, whether this be water, air, or other substance, can be held at any one temperature by a source of heating or cooling which, in turn, can be controlled by a suitable valve or switch. A further requirement is that the amount of heating or cooling available is sufficient to meet the temperature required.

As a simple illustration, if it is required to heat a domestic hot-water cylinder of some 140 litre to a temperature of 60°C, there is an obvious requirement for heating (through an electric immersion heater for instance) capable of providing sufficient heat input to raise the water temperature initially from the lowest ambient likely to be experienced up to the desired temperature of 60°C. In present-day terminology, the water in the tank would be termed the *controlled medium*, whilst the temperature at which the water is to be held would be termed the *controlled condition*. The heat input into the water would be termed the *manipulated variable*, on which the thermostatic control would operate to produce the desired result in water temperature.

Before we discuss the various possible methods of obtaining thermostatic control, we must consider a number of definitions and conditions which are more or less common to all thermostatic control systems.

Control point

This is the temperature at which the controlled medium is to be maintained.

Set-point

This is the point of the temperature scale of the thermostat to which the temperature-selection indicator is set. It may differ from the control point owing to temperature gradients within the controlled medium.

Differential

This is a term normally applicable to electric thermostats offering on-off or two-step control. The term *differential* generally represents that number of degrees of temperature change required to move the thermostatic switch from the 'off' to 'on' position; thus a room thermostat which cuts out on a rise in temperature to 18°C and cuts in when the temperature has fallen to 17°C is said to have a differential of 1°C. It should be noted that the differential declared by the manufacturer is generally measured under conditions where the change in temperature is at sufficiently slow a rate to eliminate any effect of the thermal mass of the thermostat and its possible influence on the final differential. Thus in the thermostatic air switch just mentioned if the bi-metal or thermal bellows in the thermostat is registering 18°C, and is fully saturated at this temperature, a fall in air temperature surrounding it will not immediately cause a change in temperature in the thermal element.

The makers of the thermostats will have taken account of this point both in checking and declaring the differential of the thermostat, which they will have tested under a rate of change of temperature sufficiently slow to ensure that they have measured the true differential unaffected by any lag caused by the thermal mass of the element. This point must also be borne in mind when the differential of a thermostat is checked *in situ*. Differential should not be confused with calibration tolerances, which are declared as a plus and minus figure at a certain point in the range of the thermostat.

Proportional band or throttling curve

This is applicable to thermally-operated controllers, such as valves, damper units, and torque arms which move progressively and in relation to a predetermined change in temperature. In such a system the controller assumes a certain position for a certain temperature. The proportional band or throttling curve represents the temperature change required to move the controller from one extreme position to the other. Sometimes it is declared as a percentage of the total range of the thermostatic control. Thus, a thermal device moving a valve and having a total range of 55°C could have a proportional band of 11°C, which some makers would express as 20 per cent.

Hunting

This is a rythmic change in the temperature of the controlled medium on either side of a desired control point. Many factors can contribute towards this phenomenon, among them being a lag in some part of the control cycle, or excessive correction by the control system which over-rectifies the temperature deviation resulting in this rythmic change in temperature either side of a desired control point. It could occur, for instance, in an application where a thermally operated steam valve was fitted on to a steam supply which, in turn, was applied to a water tank with the object of controlling the temperature of the water passing through the tank. If the steam input was greatly over-rated for the amount of heating required, every time the thermostat called for heat and opened the valve the inrush of steam would drive the temperature of the water upwards so rapidly that the thermostat sensing element would lag behind this rise in temperature and only close the valve some time after the control temperature had in fact been passed. This would have the effect of causing the temperature of the water to be driven considerably higher than the control point. Similarly if the draw-off of water from the tank for its industrial or process usage was done very rapidly the replacement water coming into the tank at a low temperature would cause the tank temperature to drop so rapidly that the sensing element would again lag behind the change in temperature and only call for steam after the water temperature had dropped considerably below the control point. These two separate factors, i.e. excessive heat input on a fall in temperature coupled with rapid changes in temperature of the controlled medium and a relatively insensitive sensing element have the cumulative effect of producing a continuous rythmic change which might go on indefinitely, dependent on the rate of draw-off from the water tank. Other causes of hunting can be found in a controlled circuit and in the manipulated variable and these are discussed at some length in other sections of this book.

Lag

This, the effect of slow reaction of some part of the controlled system to temperature change, can occur in varying degrees in each section of the control system.

The thermostat has its own temperature lag. For instance, sudden changes in air temperature around a bi-metal room thermostat cause a change of temperature in the bi-metal. The bi-metal, having a specific heat considerably greater than the air, will react more slowly, and (if the air temperature is falling rapidly) by the time the bi-metal has reacted sufficiently to switch the thermostat to the 'heat on' position, the air temperature will be considerably lower than the desired control point.

In a vapour-pressure system, any change in temperature of the controlled medium has to pass through the wall thickness of the sensing phial (which may also be housed in a bulb pocket). This also causes a temperature lag. The total temperature lag between the thermal system and the controlled variable can, therefore, be considerable, especially if the controlled variable tends to change temperature rapidly.

The control mechanism within the thermostat has its own lag also. In a vapour-pressure system driving a direct-acting thermostatic valve, some pressure increase in the thermal system will have to be generated before it can overcome the pressure of the steam or hot water passing through the valve. Other lags can occur in the heating or cooling medium which is attempting to correct the temperature change. Heating coils have their own inertia, refrigeration coils have a time lag in achieving maximum cooling, and electric coil heaters take some time before they are at their maximum output of heat. All these effects can combine to cause an appreciable lag between the time when the controlled medium changes temperature and the time when the correcting action reaches its maximum effectiveness.

Control mechanisms

Most of the primary sensing elements produce very small temperature reactions, so that, if accurate control is required over a narrow temperature range, the amount of direct movement or other usable effect available from the element will be very small. If, therefore, a considerable amount of work has to be done in moving heavy switch gear or opening or closing large valves against out-of-balance pressures of steam or water, some means of amplification is required. In small heating installations, such as domestic electric circuits or central-heating circuits, the magnification is of movement only and is comparatively simple. An electric switch capable of controlling 15 A can be operated by a vapour-pressure system direct, or a bi-metallic strip will achieve the same results. In a similar manner, a hydraulic expansion system can be made to operate small gas valves or steam valves direct.

In larger industrial and commercial applications, the problem becomes more complex because amplification of power as well as of movement is required, and in these instances the thermal system generally operates servo electric devices through relays and motorized valves. In more complex control circuits, the thermal element may be a resistance element or thermocouple, the signal from which is electronically amplified and then transferred to motor-operated valves and relay valves.

Offset

This is a continuous deviation of the controlled temperature below or above the value of the set-point or control point, and generally occurs in proportional control.

TYPES OF CONTROL

Two-position or on-off control

Control systems of this type are so arranged that the thermostat is in the off or the on position with no intermediate control on the manipulated variable. This is control in its simplest form as found in thermostatic control of the domestic refrigerator or electric convector heater. It can also be found in some industrial processes such as the control of vat temperatures where some variation in the controlled temperature between the on and off points of the thermostat can be tolerated.

The temperature difference between the on and off position, which is known as the differential, varies with the type of thermostat being used. In an electrically operated thermostat in which the thermal element moves the switch directly, the contacts will be gradually brought together, remain so whilst the thermostat is calling for heat, and then gradually move apart as the controlled medium reaches the desired temperature. In such cases the differential will be comparatively small. Such systems, however, are limited to fairly low electrical loadings, as high loads would cause arcing if contact were made or broken slowly. In thermostats where the slow movement of the thermal system is converted to a snap-action the differential is likely to increase owing to the necessity of pre-loading the snap device.

Because of the inherent cycling element in on/off controls they are normally used only in control systems where there is little or no lag in the heating or cooling response to the call from the thermostat. Obviously, if to the differential a considerable system lag has to be added, the final variation on each side of the desired controlled position will be unacceptable. For instance, whilst a two-position control would be quite satisfactory in the control of convector heaters sited within a room, it would not be suitable for a larger installation controlling an air-heating system for a whole building, for, by the time a falling temperature had been sensed by the thermostat, the building temperature would already be falling below the cut-in point of the thermostat. In addition, even when the thermostat had switched on the heating system, the heat exchanger would first have to warm up to reach its full output, and a further drop in room temperature would be experienced before the heat exchanger was able to bring the temperature up to the desired value. Having restored the correct air temperature in the

building, the thermostat would be lagging behind the rising temperature; thus by the time it reached the point where it should cut out,
the room temperature itself would be well above the control point.
Moreover, when the thermostat had cut out, some residual heat from
the heat exchanger would have to be disposed of, and a further rise in
room temperature would be experienced before it began to fall back.
Thus a simple on-off thermostat is always working appreciably behind
the room temperature and never has a chance to catch it up. Various
methods are available to counteract this.

On-off control with accelerator

The basic disadvantage of the two-position control just described is
that once it starts calling for heat it goes on doing so continuously
until the thermostat is satisfied, by which time the actual temperature
of the process or condition being controlled may have exceeded the
temperature of the sensing element and may reach a value well in excess
of that desired before the thermostat reacts sufficiently to cut off
the heat input. It would, therefore, be an advantage to arrange for heat
input to be administered at a rate compatible with the rate of response
of the thermal element itself, and a method of doing this has been
successfully applied in many two-position systems. When the thermostat calls for heat, this is not switched on continuously, but is supplied
in small amounts with a pause between each addition to enable the
resultant temperature rise to be transferred to the thermostat. This
keeps the heat input roughly in line with the sensitivity of the thermostat
and its ability to adjust to the increase in temperature. Thus, when the
temperature of the heated medium finally reaches the desired point,
the thermostat element will also have reached the control point and will
cut off the source of heat.

In this type of two-position thermostat extra heat to bias the temperature sensing element is introduced by a small auxiliary heater which is
built into the thermostat housing and comes into operation whenever
the thermostat is in the on position. When the air temperature passing
across the sensing element is well below the control point, it will take
longer periods of heating from the auxiliary or accelerator heater to
move the thermostat to the off position, but as the air passing across
the element increases in temperature, then the added effect of the
accelerator will cause the thermostat to switch to increasingly longer
periods of 'off' and finally to reach the control point or the set-point
where it will stay off, and remain there until the temperature has again
dropped to the point where the thermostat cuts in.

The intermittent heating effect can also be produced by rotating

cams. In such a case the cam is rotated at a fixed speed and any on-off
period can be selected by the use of movable contacts, the possible
limitation being that this on-off period, once fixed, will remain constant,
regardless of the heat-load ratio. In other words, the frequency of
auxiliary switching will remain fixed, no matter whether the thermostat
is calling for temperature correction several degrees below the control
point, or whether it is just about to reach the control point.

Proportional or modulating control

The two terms proportional and modulating control have in the past
been used to signify the same effect. In present day usage the term
proportional control is used for preference where the thermal action

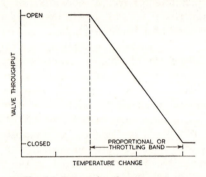

Fig. 6.1. Proportional movement.

proportionately controls both the valve and the flow through the valve
in direct relationship to change in temperature. To achieve this the
valve ports are characterized so that a truly proportioning action of
the flow through the valve is achieved. The term *modulating* action
tends nowadays to be used to describe control that provides some
proportional effect but, due to the use of valves that are not fully
characterized, produces a flow change which is not necessarily linear in
relation to a change of condition in the controlled medium, or in a
change in signal from the temperature or other type of controller.

This type of control is applicable to thermostatic systems which,
either directly or through an electric circuit, control a valve in such a
way that the valve moves to a predetermined position for a given
temperature (Fig. 6.1). The thermostat, once set to a certain temperature, opens or closes the valve over a certain throttling band. For
instance, a direct-acting valve whose stem is being operated by the
thermal system, once set so that it will be fully closed at, say 18°C

could possibly have a throttling band where it would need a change in temperature to 12°C to achieve the fully-open position; thus at a temperature of 15°C the valve would be half-way open. Let us assume also that the valve is being used on a heating system to heat a room for which the normal heat requirement is 63 MJ/h to maintain the temperature at 15°C. Finally assume that this quantity of heat corresponds to the throughput of the valve when half open. Then if the temperature tended to fall, the valve would open proportionately to the fall in temperature, admitting more hot water until the room temperature was restored to 15°C, when the valve would return to the mid-way position, and its normal throughput of 63 MJ would maintain the correct temperature. A disadvantage occurs when the heat load in the room changes drastically. If, for example, there was a sudden drop in temperature outside the building and the heat requirement to maintain

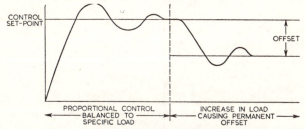

CONTROL SET-POINT

OFFSET

PROPORTIONAL CONTROL
BALANCED TO
SPECIFIC LOAD

INCREASE IN LOAD
CAUSING PERMANENT
OFFSET

Fig. 6.2. Diagrammatic representation of proportional control and the effect of a change in load.

15°C in the room rose to some 95 MJ, then the temperature in the room would drop to a new control level (Fig. 6.2) for the following reasons. The thermal element would sense the drop in temperature and open the valve to admit more heat. However, as soon as the room temperature tended to rise to 15°C, the valve would attempt to move to the mid-way position, where the heat input would again be reduced to 63 MJ. Eventually, the valve would settle at a position slightly above the half-way position but not open enough to supply the 95 MJ required to maintain 15°C; and the room temperature would permanently be offset to some figure lower than 15°C, until the heat requirement to maintain this temperature was again reduced to 63 MJ.

   The above illustration refers to a direct-action thermostat, where the thermal element is in direct contact with the valve. It is valid also for remotely operated valves where the thermostat transfers an electric signal to a modulating valve through a potentiometric balance circuit or similar arrangement.

Integral or floating control

In this system the final control moves at a slow rate in the direction desired until the thermostat is satisfied. Such a system could comprise an electric on-off thermostat which, when it calls for heat, enables a valve to travel towards the open position. However, the call for heat is interrupted through a rotating cam which only gives a positive signal to the valve during prescribed intervals. Thus the valve, instead of travelling to the fully-open position on a call for heat, moves in steps towards the fully-open position, and so enables the heating medium to put heat into the system at a rate which the thermostat is capable of sensing and keeping up with. This it will continue to do until the thermostat is satisfied. At this point there are two alternatives: either the thermostat goes to

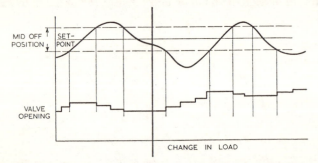

Fig. 6.3. *Diagrammatic representation of floating control. Note progressive opening of valve on a change of load condition when valve commences to control at an increased opening to maintain set-point temperatures after an increase in load.*

a reverse contact where the valve will then start moving back towards the closed position; or the thermostat may have a mid-off position, that is, a small temperature band where it is satisfied and does not send a signal to the valve either to open or to close (see Fig. 6.3). The valve will then stay in the position at which it was last set until a temperature change occurs at the thermostat, which will then send a signal to the valve, moving it open or closed as required. The system described is the simplest form of this type of control where, once the thermostat calls for heat, the valve is in operation for a constant interval in each minute. However, further refinements are possible in which the speed of the motor, or the number of seconds it is energized in any one minute, can be varied in relation to the temperature change. Thus, for a small deviation from the control temperature, the valve is moved forward or backward at a slow rate; but a sudden change of several degrees in temperature will, through suitable mechanisms, move the valve more rapidly. This enables the control system to react more quickly to conditions that change drastically. There is an obvious limitation in that the change in heat input must not be so rapid that the change in temperature of the controlled medium occurs at a higher rate than the thermostat can assimilate, otherwise continuous hunting could easily ensue.

Proportional-plus-floating control, also known as (proportional-plus-reset) control or (proportional-plus-integral) control

It has been seen that where proportional control is employed a drastic change in heat requirement to maintain a control point in the

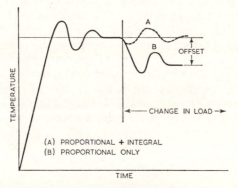

Fig. 6.4. Diagrammatic illustration of the effect of a change in load on A proportional-plus-integral control and B proportional only.

controlled medium results in a permanent offset; in other words the temperature of the controlled medium is maintained at a higher or lower value than the set-point on the thermostat. If, then, we arrange a combination of both proportional control and floating control the degree of offset can either be eliminated or considerably reduced (see Fig. 6.4). In this type of controller there are generally two adjustments, one sets the proportional band and governs the temperature change at the thermostat necessary to move the valve from the fully-open to the fully-closed position; the other control regulates the degree of floating action that can be imposed on the valve over this proportional band. With this combination, the floating or reset control will tend to correct the offset which normally would have been part of the proportional-control action.

Derivative control

Proportional control, integral control, or a combination of the two are generally acceptable for most thermostatic and automatic-control systems found in commercial use. Heating and ventilating systems and simple plant processes can use one of them to provide adequate and accurate control. However, in some processes the lag in the response to the thermostat signals and the slow rate of recovery in the controlled medium can cause either continuous oscillation or changing temperature in the controlled medium over a long period before stabilized conditions are again resumed. To overcome these disadvantages, derivative control is superimposed on one of the three controls already described. This type of control has been conveniently defined as 'a type of control in which the potential correction is proportional to the rate at which the deviation changes'. Thus, by adding derivative control to proportional control, the offset associated with proportional control still remains, but the number of cycles or oscillations through which the controlled medium will pass before stabilized conditions are achieved will be considerably reduced. When added to proportional-plus-integral or reset control, it will again effectively damp out the oscillations that occur before stabilized conditions are reached; and will still retain the advantage of normal proportional-plus-integral control in that offset will either be eliminated or considerably reduced.

7

ON-OFF OR TWO-POSITION THERMOSTATS

TWO-POSITION thermostats providing on-off control through electric contacts form by far the majority of thermal-control devices in present-day use; and within their limitations they provide the most economic and simple thermal-control systems for most applications where the temperature of a substance or condition is required to be held within certain limits. Such is the diversity of application for this type of device that the word 'thermostat' is now generally qualified with a prefix such as air-thermostat or liquid-immersion thermostat, and each of these sub-types differs in points of detail in order to suit particular applications. On-off thermostats as a whole, however, have certain basic features, and there are requirements common to all types, which will now be considered.

Thermal element

All the thermal elements previously discussed can be used for on-off control of switches, but some types lend themselves more easily to this problem and are therefore used in most controls of this type. These include vapour-pressure systems, bi-metal, and rod-and-tube elements, three types which represent probably the most economic means of obtaining thermal movement. Their low thermal mass makes them suitable for systems where the thermostat has to sense air or gas temperatures, and where a thermal element having a higher mass would suffer from excessive lag. Hydraulic systems are frequently used when remote temperature sensing is required, and when some advantage can be obtained by using a bellows or diaphragm remotely connected to a sensing bulb by means of capillary tubing. The hydraulic system is especially useful where long temperature ranges are required; in this respect it has some advantage over the vapour-pressure system with its comparatively short range length.

Ranging and transmission members

The majority of on-off thermostats are expected to provide a range of temperature selection within which they can be set to any point of temperature control. In most designs the range mechanism is incorporated in the transmission link between the thermal movement and the switch, and arranges that a positive connexion between these two members is only provided when the thermal member is sensing temperatures at the set-point. Thus the ranging mechanism must absorb or resist thermal movement up to the set-point, after which it allows the thermal mechanism to move the switch, either making or breaking circuit according to the application and the direction in which the temperature surrounding the thermal element is moving.

Switch mechanism

This can be of the creep-action type where the movement of the contacts towards each other is in direct relationship to the speed of the thermal movement; but generally some quick or snap-action movement is incorporated to avoid arcing and contact wear. As the current being controlled increases so the need for a snap-action movement also increases; eventually there comes a stage at which the thermostat contacts themselves are no longer able to control the load, and they have to operate in conjunction with a suitable relay circuit.

Whilst the standard relay meets the requirements for most applications, there are many instances where a very fine degree of control is required, and where the thermostat must have a very low thermal mass in itself and be capable of moving very slowly under a gradual temperature change. With a standard relay this would entail comparatively long periods during which the on-off contacts would be close together near the closed position, and this would often cause chatter or rapid cycling of the normal relay. This problem is especially acute when thermostats are used on inductive circuits, such as solenoid and motor windings.

Hot-wire switch

In such applications the hot-wire vacuum switch (H.V.S.) offers an acceptable solution. This depends on the fact that when an electric current is interrupted by the separation of two surfaces in a vacuum no arc is formed. In these conditions the separation of the contacts need only be of the order of 0·025 mm, so that if a switch works in a vacuum the contacts can be very light and the movement very small. The movement necessary to close or open the switch is readily obtained through the thermal expansion of a wire through which the control current is passed. The principles of the hot-wire vacuum switch are illustrated in Fig. 7.1. T and T' are terminals of the main circuit it is desired to control. A and B are fixed conductors, A terminating in the spring S attached to level C. This spring tends to swing the level about the fixed fulcrum F, closing the gap G. At the other end of level C is bobbin D, round which passes the resistance wire W, rigidly fixed to the terminals H and H'. By tensioning this wire, spring S is compressed,

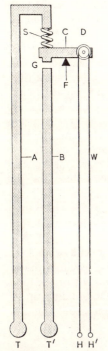

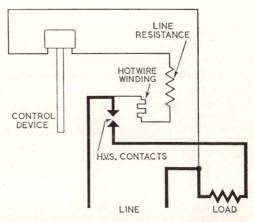

Fig. 7.2. *Typical application for a hot-wire relay. The wiring is arranged for control by a thermostat which opens on reaching a predetermined temperature. Here the opening of the control circuit through the H.V.S. allows the main contacts to open and cut out the heating load.*

Fig. 7.1. *Principle of hot-wire vacuum switch.*

(*G.E.C.—Sunvic.*)

gap G opens and the switch is in the open or off position. If a small current is passed through wire W, it is heated and expands, allowing spring S to force lever C down, and throw the switch into the on position by closing the gap G.

The winding W consists of a number of turns of special steel wire wound between two Steatite insulating bobbins. The temperature when the winding is energized is well below that at which creep is liable to occur, and the properties of the winding are not affected by frequent operation over long periods of time or by its being continuously energized. The current in the winding varies from about 25 mA in the smallest switch of this type to 70 mA in the largest. This type of relay lends itself to many applications where a fine degree of control is required, and yet where the temperature sensing device is only capable of handling a very small current. Fig. 7.2 shows how a hot-wire vacuum switch can be wired to operate a thermostat.

Almost any control device can be used with this switch, fitted with simple contacts such as are found on contact thermometers or bi-metal switches. The low non-inductive load presented by the hot-wire winding enables a.c. or d.c. supplies to be used at line voltages without danger of arcing at the control contacts. Another advantageous feature of this type of relay is that when the primary control circuit is made, some time is taken by the hot-wire winding to heat up and make the load circuit, and in this respect the relay is not instantaneous in action. This prevents chatter at the load-carrying switches. Where time delay is a requisite part of a larger control circuit it can be increased by the manufacturer of the relay to times up to 20 s within certain tolerances.

With this combination thermostatic control can be provided giving on-off control over electrical loads up to 25 A at line voltage and yet provide a thermostat differential in the order of $0 \cdot 1 °$C. Obviously to this last figure must be added the effect of load and heating lags which may produce a final cycling on either side of a mean figure somewhat larger than this.

As previously stated, the hot-wire vacuum switch relay can be used for switching inductive loads, but for this application it is essential that a surge suppressor be fitted across either the H.V.S. contacts or the load contacts. This precaution avoids the possibility of the building up of high transient voltages by the stored energy in such circuits at the moment of break.

Differential

As already stated, this is the temperature change to which the thermostat must be subjected before its contacts will move from one position to the other. All thermostat systems suffer from a degree of *hysteresis* whereby for a certain small change in temperature there will be no movement of the thermal mechanism, or of the complete system. To keep hysteresis as low as possible the thermal mechanism should be as sensitive as the application permits, and the linkage between the thermal element and the switch should be carefully designed to keep lost motion down to as low a figure as possible. In addition to hysteresis proper, some temperature change is required to enable the thermal movement to transmit sufficient energy to the switch to overcome its snap-action device whether this be by magnet or spring. These two factors, hysteresis and minimum energy to operate the switch, govern the minimum differential of on-off thermostats. Even if it were possible to eliminate these factors so that the thermostat was able to operate virtually without differential some artificial means of introducing this would be necessary, for in most applications it would be impossible, or undesirable, to have the thermostat continuously cycling over an infinitely small change in temperature. Therefore most thermostats incorporate a controlled differential adjustment. In some cases this is set by the manufacturer, whilst in others provision is made for customer adjustment within certain limits. It is important to remember that the differential as declared by the manufacturer is a performance of the thermostat measured under controlled rates of temperature rise, in most cases being calculated when the thermostat is subjected to a rate of change of temperature of less than $1°C/min$. In service conditions, a thermostat may be subjected to more rapid changes of temperature; and some account must be taken of the lag caused by the thermal mass of the sensing element and the thermostat as a whole, which can have the effect of increasing the apparent differential of the thermostat.

Thermal lag

No thermostat is required to switch on and off without also controlling some form of heating or cooling; and thermal lag or hunt can only be properly assessed if the effects of both the thermostat and the source of heat are considered at the same time. As already stated, the thermostat has a thermal mass of its own and the higher this is the more likely is the thermostat to lag behind the temperature changes in the air or liquid surrounding it. The source of heating or cooling being controlled by the thermostat will have a similar lag, caused by its own thermal mass, so that this too will be more or less sluggish in its ability both to provide heat immediately it is called for and to stop providing heat as soon as the thermostat switches it off. Therefore, if the thermal lag of the thermostat and of the heat input are working together the final effect on the temperature of the medium being controlled could be fluctuating temperatures either side of the control point. The total effect of lag is known as cycling or hunting. This will be most severe when the thermostat and the heat source have high thermal lag.

There are various methods of reducing thermal lag in the thermostat, the obvious one being to reduce the thermal mass of the thermostat. But the extent to which this can be done is limited because the thermal element must do a certain amount of work.

Thermostat switching

When large loads are controlled by a thermostat having a very fine differential the use of hot-wire vacuum switches is an advantage. In domestic thermostats it is becoming the practice to use low-voltage circuits of the order of 24 V. This, among other things, offers the advantage of thermostats with finer differential and lower thermal mass and makes for the achievement of finer control generally.

The snap-action of the switch movement is generally obtained by either an over-top-dead-centre spring arrangement, or a magnetic pull (Fig. 7.3). These have been described in some detail in Chapter 5.

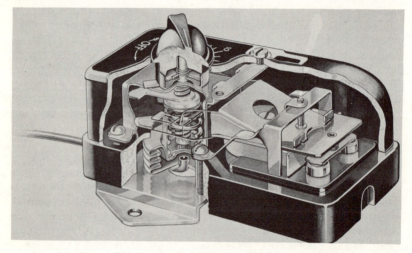

Fig. 7.3. Commercial thermostat with magnetic snap-action contact.
(Teddington Autocontrols Ltd.)

Their purpose is to enable the contact to be brought together at a greater speed than the comparatively slow thermal movement would normally permit. This, of course, can only be achieved by some increase in differential.

In addition to these open-contact type switches, where the contact mechanism is virtually an integral part of the transmission, there are other types where the contact mechanism is totally enclosed. One type is the mercury tube switch. In this two electrodes are sealed within a glass tube which also contains a quantity of mercury. When the tube is tilted in one direction the mercury is clear of both electrodes, but a tilt in the opposite direction moves the mercury to a position where it completes the circuit between the two electrodes. Many versions of the basic mercury tube are available, whereby a simple thermal movement in relation to temperature can achieve a multiplicity of switching operations.

The totally-enclosed mechanical switch is becoming of increasing importance in thermostatic control and forms the basis of many thermostats in current use. In these the switches are of the micro-break type in which all the intermediate levers and moving parts are as small as possible and are housed in moulded cases which are virtually dust proof. The only parts normally visible are the terminals and a plunger or lever to which the thermal movement is applied to achieve switching.

Fig. 7.4. Typical room thermostat for domestic and commercial use. (Satchwell Controls Ltd.)

Air thermostats

These are designed primarily to act as temperature controllers in rooms and buildings and, by on-off switching, are expected to control electrically a source of heat sufficiently accurately to provide comfortable conditions for the people in the room (Fig. 7.4).

Two types of thermal element are used in the majority of air thermostats, namely vapour-pressure systems and bi-metal. The vapour-pressure system is especially useful where a strong thermal movement is required, e.g. for thermostats directly controlling electric switches for loadings up to 3 and 4 kW. They generally consist of a self-contained bellows partially filled with a suitable liquid to cover the range of normal air-temperature control. To provide maximum sensitivity the bellows must be extremely flexible; this is achieved by using bellows of very thin wall thicknesses and the deepest possible convolution. In addition the bellows dimensions are such that they have the lowest possible spring rate and are, therefore, able to react to the slightest change in vapour pressure.

The majority of air thermostats, whether they use vapour-pressure or bi-metallic thermal systems, incorporate a snap-action device. However, some bi-metal thermostats are designed so that the gradual movement of the sensing blade is transferred indirectly to the switch. The switch arms have a mechanical link joining them to the bi-metal. This link incorporates a lost motion device which ensures that at the instance of break the bi-metal is already moving and accelerating in its movement at the point where it imparts movement to the switch arms. In this manner adequate contact pressure is maintained right up to the moment of break, and then a clean break is achieved without arcing. Thermostats incorporating this arrangement are available capable of handling loads of up to 5 kW at 250 V.

Other types of bi-metal, taking the U shape or helical winding, are also used as air thermostats. These have already been discussed in Chapter 1.

Accelerator thermostats

As mentioned earlier, a basic problem in the use of the on-off thermostat is that because of its thermal mass the thermostat will always lag behind the room conditions at any given time. Consider a thermostat in a room where the temperature has been static for some hours. Under these conditions, the bi-metal or vapour-pressure system will be exactly at room temperature; but if the room temperature then starts to rise the thermostat must immediately begin to lag behind because it will be necessary for the air-temperature change to be transferred through the thickness of the bi-metal or the liquid in the vapour system. Obviously, the more rapidly the room temperature changes then the greater will be the lag of the thermostat.

In Chapter 6 it was shown that this problem of thermal lag and thermal inertia could be largely overcome by introducing into the thermostat a small electric heater which, whenever the room thermostat calls for heat, is switched on and imparts heat to the sensing element in addition to the heat it receives from the rising air temperature round it. The methods of wiring the heat accelerator are shown in Fig. 7.5. In the *series* arrangements the heater in the thermostat is in series with the main load. Being of low resistance it provides sufficient heat for the purpose of heating the bi-metal without materially affecting the load characteristics of the thermostat. As the current consumption of the heating load varies from one application to another, and this materially affects the heat output required from the accelerator, the manufacturer provides a means of adjusting the system (the heating load control) to ensure that it provides the heat required.

In the *shunt* accelerator heater arrangements (Fig. 7.5(c–e)) the

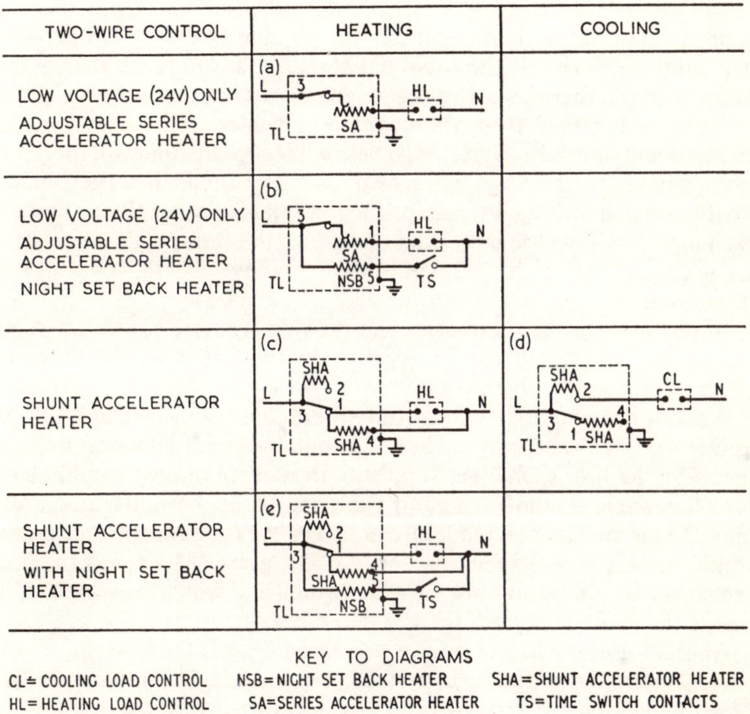

TWO-WIRE CONTROL	HEATING	COOLING
LOW VOLTAGE (24V) ONLY ADJUSTABLE SERIES ACCELERATOR HEATER	(a)	
LOW VOLTAGE (24V) ONLY ADJUSTABLE SERIES ACCELERATOR HEATER NIGHT SET BACK HEATER	(b)	
SHUNT ACCELERATOR HEATER	(c)	(d)
SHUNT ACCELERATOR HEATER WITH NIGHT SET BACK HEATER	(e)	

KEY TO DIAGRAMS

CL = COOLING LOAD CONTROL	NSB = NIGHT SET BACK HEATER	SHA = SHUNT ACCELERATOR HEATER
HL = HEATING LOAD CONTROL	SA = SERIES ACCELERATOR HEATER	TS = TIME SWITCH CONTACTS

Fig. 7.5. Various wiring methods for air thermostats using accelerators. (Satchwell Controls Ltd.)

heater is in parallel with the main heating load. This has the primary advantage that the heater output is constant regardless of changes in any other electrical demands being made through the thermostat; but, on the other hand, it is virtually non-adjustable, so that a new heater is sometimes required if the characteristics of the thermostat have to be altered in any way.

Operation of accelerator thermostat

The operation of the accelerator thermostat is as follows. As soon as the room temperature drops the thermostat makes circuit and calls in the heating load. At the same time it receives heat from the accelerator so that it is now no longer necessary for the sensing element to await a heat transfer from the rising air temperature in the room. Instead it receives excess heat until its temperature is restored to the control point and it opens circuit. At this point the room temperature may be below the set-point; but, since the heat exchanger supplying heat to the room has its own thermal lag, heat will continue to be put into the room even after the thermostat has cut off, and the room temperature will continue to rise. If the room temperature fails to reach the control point then the thermostat will again cut in as soon as the bi-metal has lost the heat gained from the accelerator heater. This time it makes circuit it will only be a degree or so below its set-point, and will therefore react quickly to the accelerator heat after the circuit has been made. In this way, it will again switch off quite quickly and allow the heat exchanger to discharge its residual heat into the room without creating an overshoot when the temperature rises above the set-point of the thermostat.

In this manner the thermostat can provide accurate control of room conditions, and avoid the excessive overshoots that would normally occur with a simple thermostat.

A severe drop in room temperature would cause the accelerator to take longer to heat the sensing element up to its set-point temperature, and therefore the heat exchanger supplying the source of heat would be on for a long spell. On the other hand, for a small drop in room temperature once the thermostat had made circuit and called for heat, the accelerator would quickly heat up the element to its set-point. As a result the heat exchanger would be on only for a short time and would be switched off before the room temperature exceeded the set-point.

Another use of a heater built into a thermostat is the lowering of the control point for a specified period, as during night running. In this type of switching, a second heater is introduced into the thermostat housing and this is only in circuit during certain hours of the night. With the night heater energized the sensing element in the thermostat is subject to a continuous source of false heat and therefore will only make circuit when, in spite of this false heat, the air temperature drops sufficiently to cause the sensing element to make circuit. This has the total effect of down setting the control point on the thermostat by some predetermined amount, generally 5°C. Wiring of night set back heaters is shown in Figs. 7.5(b) and (e).

Night offset controls, and time switches

In addition to the above example there are several other arrangements whereby temperature control in a building or room can be auto-matically set down through the night period. Each of these combinations must include a clock to control the time of operation, and this can either be an integral part of the thermostat or can be remotely sited. A method using a remote clock is shown in Fig. 7.6 where two thermostats are employed, one set to the required day-temperature control point, and the other to the night-temperature control point.

The electric clock has changeover circuits, or some other arrange-ment, so that on night control the day-temperature thermostat is shorted out and the night-temperature thermostat is in control; in the morning the reverse takes place. The two thermostats can either be completely separate or can be built into one control box with provision for scale and set-point observation. The use of a separate clock has some advantage in that the thermostats can be sited where the most representative average temperatures in the building or room can be sensed, whilst the clock can be housed with other basic controls close to the heat exchanger, boiler, or other source of heat.

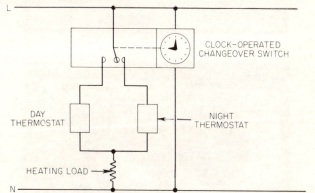

Fig. 7.6. Diagrammatic arrangement of two thermostats for day/night control using clock-operated change-over switch.

Other thermostats are available with a built-in clock, the one shown in Fig. 7.7 being a typical example. In this arrangement there is one thermal element with a manual range-setting mechanism. The clock is driven by an electric motor and, in addition to carrying the normal clock-face hands, is arranged to drive a cam which bears on and (during the night) over-rides the manual adjustment of the set-point on the thermostat. In this way, during day-time operation the thermostat operates normally, controlling at the set-point selected by the manual scale adjustment. When the point of night operation is reached, the cam that is being driven by the clock mechanism comes to bear on the range-setting mechanism and over-rides the set-point selected by the manual pointer. According to the characteristics of the cam, the set-point is thus set down to the temperature required for night running. This condition is maintained until morning when the cam relinquishes its bias on the range setting and the thermostat returns to control at the set-point prescribed by the manual adjustment.

Heavy-duty industrial air thermostats

Whilst the controls so far described can be supplied in protective

Fig. 7.7. Room thermostat with built-in clock and clock control for day/night setback. (Honeywell Controls Ltd.)

covers for commercial use there are, in addition, air thermostats sufficiently robust for heavy-duty industrial applications where some corrosion has to be resisted. Of these, the one shown in Fig. 7.8 is typical. The thermal system is a vapour-pressure one consisting of a self-contained bellows housed in the cup on the top of the die-cast case.

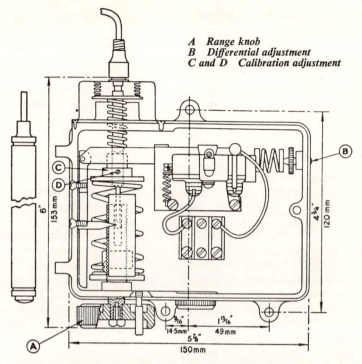

A   Range knob
B   Differential adjustment
C and D   Calibration adjustment

Fig. 7.8. Commercial thermostat for air temperature control, shown with sensing phial and capillary. (Teddington Autocontrols Ltd.)

The bellows is on a common centre-line with the range mechanism, which consists of a spring, whose compression can be varied by rotation of the set-point adjuster at the base of the diecasting. Rotation of this adjustment increases or decreases the opposing spring force on the bellows, and, at the same time, moves a scale marked in degrees of temperature against a fixed pointer on the side of the case so that indication of set-point is provided.

On a fall in temperature the bellows recedes against the spring force and allows the spindle connexion between the spring and the bellows to move upwards. This connects with a lever pivoted on the left-hand side of the case which, in its travel, tilts a mercury tube pivoted in the centre of the thermostat case. The movement of the lever is partially restricted by a snap spring located between the end of the lever and the right-hand side of the case. The compression of this spring is adjustable so that the differential of the thermostat can be set within certain limits. The use of a mercury tube offers considerable advantages when the thermostat is called upon to operate in industrial atmospheres where open contacts would suffer from high humidity, corrosion, and general impurities.

The same type of thermostat can provide remote control by extension of the thermal system to include a sensing phial and connecting capillary. In this manner the thermostat can be made to operate on outside air temperature or ducting temperature, whilst still retaining the ability to position the thermostatic switch inside the building in closer proximity to other controls.

Air thermostats for proportional and floating control

Although this chapter is primarily concerned with on-off control it might be useful to draw attention to the fact that the same basic controls can be modified to provide either proportional or floating control. In these instances the on-off switching is replaced by other devices. In electric proportional control the thermal system transmits movement through a transmission and ranging mechanism to a potentiometer. Provision is generally made for the proportional band to be adjustable over a percentage of the total temperature range. In these systems, the thermostat controls the position of an actuator, such as a valve or motorized damper unit, and can modulate the position of these any stage from full on to full off depending on the temperature deviation from the set-point.

When used for floating or integral control, the thermostat is designed to operate a pair of changeover switches; in some cases this is increased to two pairs of changeover switches to provide two speeds of response in relation to temperature change. The thermostat is provided with a mid-off position so that when the thermostat is satisfied no circuits are made and the system is at rest. Should the temperature rise or fall a small amount in relation to the set-point, the thermostat will make one of the first sets of contacts which will set in motion the motorized valve or damper unit towards the position which will cause a temperature correction. On this first circuit the motor will be energized for a few seconds only in every minute, by the action of a rotating cam in the control circuit. If a further deviation in temperature takes place, the thermostat will move beyond the first set of contacts to a second set which will short out the cam and cause the valve motor to run continuously. As the temperature of the air being measured moves back towards the control point the thermostat will break the continuous-running circuit and return to the first circuit, causing the motor to continue towards its correcting position but at the reduced speed. Finally as the air temperature reaches set-point the thermostat will return to the mid-off position and put the whole circuit to the 'at rest' position, until a further temperature deviation takes place. The motor, being a reversing type, can move in either direction, and move the valve or motorized damper unit towards the fully-open or fully-closed positions in accordance with requirements.

Air thermostat location

Most air thermostats are used for the control of room temperatures. They are normally provided with a temperature range selection which relates to room temperatures, and it is therefore desirable that their location enables them to sample true room conditions. Where this is impossible to achieve it is necessary to locate the thermostat at a point where it is set at a relative figure that bears some relation to comfort elsewhere. However, any relative control brings with it problems of maintaining a constant relationship between actual temperatures in the occupied zone and the temperatures on which the thermostat is working. Hence, wherever possible, the thermostat should be sited in such a position that it samples the temperature conditions in the part of the building for which effective control is required. The primary consideration, therefore, is to protect the thermostat from any extraneous source of heat or cold that could prevent it from operating under its ideal condition.

In domestic heating systems, where one thermostat is called upon to supply the heating control for the whole of the house or flat, it is sometimes very difficult to find a position in the house which provides the ideal control, for wherever it is sited it cannot take into account all the factors of heat losses that will apply in the different rooms within the heating area. At best an average position can be found in the hall or living-room, as representing the most used parts of the house. In the hall, there can be extraneous sources of cold, such as draughts through doors and windows, which may cause the thermostat to operate for periods of such length as to cause overheating in the rooms having lower heat losses. On the other hand, a thermostat sited in a living-room may receive extraneous heat from additional heating or from the occupants of the room, so that it may prematurely shut off the heating system to the detriment of the rest of the house.

In general several 'do's and don'ts' apply, some of the most important of which are as follows:

(1) Always mount the thermostat on an inside wall in preference to an outside wall, and if possible provide an air gap between the thermostat and the wall immediately behind it.
(2) Choose a position where air circulation is good: avoid nooks and recesses where air movement is restricted.

(3) Avoid chimney breasts where extraneous heat can be obtained from coal fires in other parts of the building, and positions over reading lamps, television sets, etc., where there is extraneous heat rising.
(4) Avoid mounting the thermostat where it can receive radiant heat from radiators or low-temperature panels.
(5) Finally, avoid mounting the thermostat where it can receive direct sun rays through windows.

There will be other considerations which dictate the final position, such as the run of the wiring and the ability to conceal this, so that in the end the final position selected will be a compromise between these many factors. Nevertheless, it must be borne in mind that the mounting of the thermostat is an extremely important consideration; and experience has shown that if these factors are disregarded an otherwise satisfactory heating system and thermostat can easily be condemned by a dissatisfied user.

Immersion thermostats

Sometimes known as aquastats or duct thermostats, these devices offer two-position thermostatic control for air or gases in ducting, and temperature control of liquids in enclosed vessels or vats, ranging from domestic to industrial process control. They are frequently used for switching electric heaters and fans, or for positioning such items as motorized valves, damper units, and solenoid valves. Despite their variety of application, their function as an on-off or two-position switch exhibits certain characteristics similar to those of the air thermostat.

For the most part the thermal element is arranged so that the sensing bulb is remote from the actual switch mechanism. This can be arranged either so that the bulb is external to the switch mounting but is a rigid fixing to it, or the bulb can be quite remote as in the bulb and capillary arrangement. Choice is largely governed by the type of thermal system employed.

With rod-and-tube type or bi-metal elements the switch mechanism is generally attached to the end of the sensing device so that the thermal movement can be conveyed directly to the electric switch. Fig. 1.3 shows a type of thermostat frequently employed in domestic equipment where the liquid or air being controlled is in an enclosed vessel, and where conditions permit the mounting of the thermostatic switch external to this.

For applications where switch operation must be remote from the point of temperature sensing, the vapour-pressure or hydraulic-charge thermal system, utilizing a sensing bulb connected to a bellows or diaphragm by means of capillary tubing, is the best choice; thermo-couple types are, however, also used. For general industrial applications where range selection over a small band of temperatures up to 150°C is required, the vapour-pressure system probably offers the best solution, because it provides the maximum thermal movement at the point of control. Where wide temperature bands are to be used the hydraulic system has an advantage as it is not limited to the useful pressure range of any one liquid. However, as the thermal movement is comparatively small over a few degrees of temperature change, devices utilizing this thermal system are generally limited to the control of small electrical loads, unless relays or contactors are employed.

This problem similarly effects the differential obtainable with this type of thermostat. Differential being a function of the thermal mass of the sensing element, the degree of snap action required, and the load to be carried by the switch, very fine differentials are only possible where these factors are small; it therefore invariably follows that when differentials of less than 1°C are required, and where the mains-current load exceeds approximately 5 A, a relay of some sort must be used.

In most thermostats used in ducting and platens the sensing elements are inserted into a liquid, gas, or metal, and some precautions must be taken in choosing the point of insertion. This is especially important when the medium being controlled is air or gas where the rate of heat transfer to the sensing element is comparatively slow. The essential feature is to ensure that the whole of the sensing element is subject to changes in temperature of the medium being heated, and that, wherever possible, some agitation is present to cause a flow across the sensing element. Stagnant pockets or layers of slow moving air should be avoided, and a position chosen which is as representative of true temperature as possible.

Whilst in most air applications the sensing phial can be inserted into the duct without need for special protection, where thermostatic con-trol is being effected on most gases and liquids some protection is required. This fulfils two functions: it protects the sensing element from attack by the medium being heated, and allows for the removal of the thermal element without the need for stopping the process being controlled. Most manufacturers of thermostats for this type of applica-tion supply associated bulb pockets with their controls (see Fig. 7.9). These are designed to effect a snug fit on the sensing bulb and are thin-walled in order to reduce the time-temperature lag through the walls. Nevertheless, the total time-temperature lag of the sensing bulb and its pocket can be considerable; therefore the characteristics of each type of thermal element must be considered before one is chosen for any one particular application. Obviously, where temperatures change

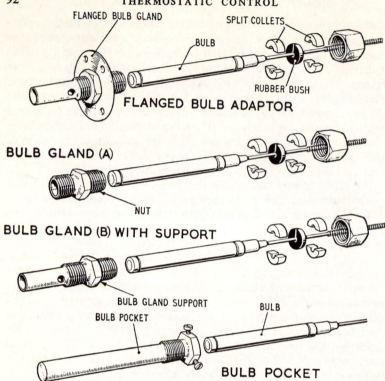

FLANGED BULB GLAND

SPLIT COLLETS

BULB

RUBBER BUSH

FLANGED BULB ADAPTOR

BULB GLAND (A)

NUT

BULB GLAND (B) WITH SUPPORT

BULB GLAND SUPPORT

BULB POCKET

BULB

BULB POCKET

Fig. 7.9. Various types of bulb pocket for duct mounting, etc. (Teddington Autocontrols Ltd.)

slowly, most thermal systems will be able to follow the trends and apply the correcting force; but where temperature conditions are likely to change rapidly and call for rapid response from the thermostat then a thermal system having the necessary speed of response must be chosen.

Surface-contact thermostats

Instruments of this type differ from the duct and immersion type in that they are designed to sense the temperatures on the surface of a body. They are all, therefore, to a varying degree subject to false heating or cooling by ambient temperature. Because of this, in order to maintain a body at a certain temperature they must to some extent operate at a higher or lower temperature than the body. In their simplest form they can be bi-metallic thermostats where one end of the bi-metal is clamped to the solid body.

A typical example is the thermostatic control on the domestic iron, in which a strip of bi-metal is held in firm contact with the sole plate. The heat transferred from the sole plate to the bi-metal through the majority of its length causes it to warp, and in so doing to open an electric circuit at a predetermined temperature. In spite of the good thermal contact between the sole plate and the bi-metal, there is a temperature gradient throughout the length of the bi-metal and some trial and error has to be carried out in the design stage so that the switch is broken at a temperature of the bi-metal corresponding to the desired temperature of the sole plate.

A further example of this type of control is the contact thermostat frequently used for sensing the temperatures of hot-water pipes in domestic central-heating circuits. This instrument employs either a bi-metal or a vapour-pressure bellows, either of which is at the back of the unit so that when the latter is clamped to the pipe being controlled the thermal element is in hard contact with the metal through which the temperature of water is transferred (Fig. 7.10). In some installations the thermostats, instead of being clamped to the pipes, are cold-welded to it by means of some thermo-setting solution having good thermal conductivity.

Other types of contact thermostat are available, for instance those described in Chapter 10 (domestic-appliance controls) now in use for controlling cooking operations by means of a sensitive element held in

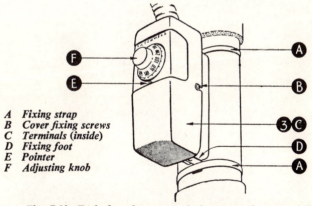

A   Fixing strap
B   Cover fixing screws
C   Terminals (inside)
D   Fixing foot
E   Pointer
F   Adjusting knob

Fig. 7.10. Typical surface-mounted thermostat for control based on surface temperature of a hot-water pipe. (Satchwell Controls Ltd.)

close contact with the base of the cooking vessel. Although these devices were primarily developed for the domestic market, they are already finding useful applications in a number of industrial and commercial processes. In these types the sensing element is a hydraulic unit connected to a bellows or diaphragm by means of a capillary. These flexible members are made to operate electric switches or gas valves and, if correctly installed, can maintain temperatures within 2–3°C.

For industrial purposes, especially where temperatures are high, the thermocouple represents the best means of effecting surface contact temperature control. It has low thermal mass, and is comparatively easily arranged to have good thermal contact with the body being controlled and to be protected from extraneous sources of heat gain or loss. The thermocouple will provide control through the medium of a direct deflexion, or through continuous balance instruments with their associated servo-mechanisms.

8

FLOATING AND
PROPORTIONAL CONTROL

FLOATING OR INTEGRAL CONTROL

For industrial plants and air-conditioning installations where the lags in temperature response of the controlled medium are small, and where the controlling thermostat is quickly able to detect not only tendencies for the controlled medium to change but also the corrective action being taken, the two-speed floating control offers a comparatively simple method of thermostatic control. Prior to the general acceptance of electronics in control systems, floating control was frequently achieved by electro mechanical systems of which the Satchwell X system was typical, and as a means of demonstrating how floating control was achieved, it is worth considering how this operated (Fig. 8.1(a)). In this system the thermostat detects the direction and the amount of the change in the temperature of the controlled medium, and then selects the appropriate direction and rate of response of the valve or damper motor. One of the two rates of response is adjustable so that it can be matched to the characteristics of the installation.

The thermostat is located in the medium which is being controlled and is set to the desired value. Connexions from the thermostat lead to the control box, which incorporates an impulse timer governing the number of seconds in each minute that the motor runs.

With the availability of electronic systems, floating control can become considerably more sensitive and need have fewer mechanical parts. In a typical system shown in Fig. 8.1(b) the temperature-sensing component takes the form of wire resistances forming one half of an a.c. bridge, the other half of the bridge comprising a set value potentiometer. In addition there is a leg of the bridge comprising a pair of temperature sensitive resistors TSR1 and TSR2 in contact with a heater winding. With no deviation the bridge is balanced, there is no output and these heaters are off.

When a deviation unbalances the bridge the output is processed for magnitude and direction, and the appropriate output relay is energized. Contacts a or a' drive the motor in the necessary direction to correct the deviation. Simultaneously, contacts b or b' complete a

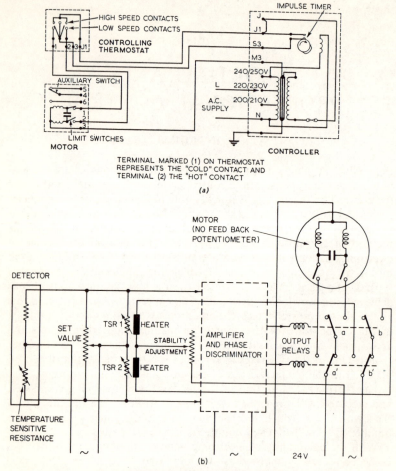

Fig. 8.1. *Floating control.* (a) *Wiring diagram for typical circuit,* (b) *electronic integral controller principle.*

circuit to one of the heaters, thus changing the resistance value of either TSR1 or TSR2, in such a sense as to re-balance the bridge by cancelling out the deviation. The output relay is de-energized, stopping the valve motor and breaking the circuit to the heater. TSR1 or TSR2 cools down and, if the deviation persists, the bridge is unbalanced once more so the process is repeated. The motor is thus started and stopped alternately, receiving a series of pulses to modify its effective running speed.

Large temperature deviations require long heater-on periods to cancel them, and the cooling is then relatively rapid. The motor therefore has a high percentage on time and a low percentage off time. As the deviation gets progressively less, due to the corrective action thus taken, so the percentage on time becomes smaller and smaller so that the motor runs slower and slower. When the deviation is zero (or within the differential of the output relays) the motor becomes stationary. There is thus no offset involved in such a system, which is its main advantage over proportional control.

PROPORTIONAL CONTROL

Electric systems

Proportional control can be provided by temperature sensing devices transmitting their information to motorized valves and damper units via a circuit which incorporates the basic Wheatstone bridge arrangement.

The Wheatstone bridge comprises a circuit basically as shown in Fig. 8.2. This consists of two sets of resistors R1, R2 and R3, R4, connected across a d.c. voltage supply. A galvanometer is connected across the parallel branches of these two sides, at the junction between the series resistors. If the circuit is completed and the resistance of the R1, R2 leg is equal to that of the R3, R4 leg, there will be no reading on the galvanometer, and the circuit is said to be in balance. However, should one of the resistors, say R1, be altered in resistance, then the current balance will be disturbed and the galvanometer will register a potential difference, indicating that the currents in the two legs are now different. This condition will persist until the resistance in the R3, R4 leg is again made equal to that in the R1, R2 leg, when the balance is again restored.

This basic principle has been applied to electric proportional control systems by several manufacturers; Fig. 8.3 shows the control circuit of a typical example.

It will be seen that resistors R1 and R2 are a single potentiometer. In most circuits this potentiometer has a value of 135Ω, and across it moves a wiper, the position of which is controlled by the thermostat. As the temperature changes the wiper moves along the potentiometer changing the values of R1 and R2. The bridge circuit is completed by a similar resistor (R3, R4) of 135Ω adjacent to the motorized valve or damper unit. This has a similar wiper which is connected to the motor shaft, so that movement of the motor moves the wiper along the potentiometer. Instead of a galvanometer in the centre of the bridge there is a relay which is sensitive to any degree of unbalance set up by movement of the wipers, and conveys a signal to its reversing or pulling motor in the motorized valve. The circuit is balanced in that R1

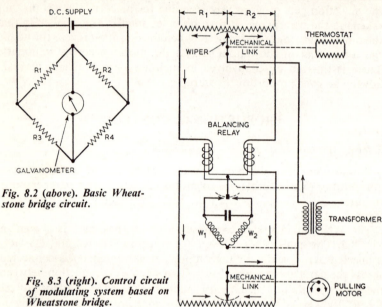

Fig. 8.2 (above). Basic Wheatstone bridge circuit.

Fig. 8.3 (right). Control circuit
of modulating system based on
Wheatstone bridge.

balances R4 and R2 balances R3. When a temperature change occurs round the thermostat the thermal element will cause the wiper on potentiometer R1, R2 to move, and upset this balance. The relay, sensitive to this changed condition, will swing one way or the other, according to the direction in which the wiper has moved, and in so doing will switch on the motor that drives the valve or the damper arm. As soon as the motor shaft begins to turn it will move the wiper on its own balancing potentiometer (R3, R4) and this movement will continue until the resistors R1, R2 and R3, R4 are again in balance. The currents in the two legs will then be equal again and the balancing relay will go to the 'at rest' position and de-energize the motor.

If the temperature round the thermostat rises, the balance in the system is disturbed and the relay switches on the motor, sending it towards the closed position. The motor, in moving towards the closed position, will also move the wiper on potentiometer R3, R4, to which it is linked. The movement of the wiper will continue until it reaches a point where R3, R4 exactly balances the values set on R1, R2 by the thermostat. At this stage both legs of the bridge have equal current flow, and the relay returns to the 'at rest' position, the motor stopping and leaving the valve at a stage nearer to the closed position. The motor can be either a low-voltage type drawing its supply from the output side of the transformer supplying the control circuit, or a mains voltage motor using the supply from the input side of the transformer through the relay contacts.

High- and low-limit thermostats

This basic circuit lends itself to many adaptations and additions. A simple variation is where an additional resistor in the form of a thermostatically operated potentiometer is put into one leg of the standard circuit. This can be used either as a high-limit or low-limit control which can to a degree over-ride the control being effected by the first thermostat. If such a controller has a standard 135Ω resistor incorporated in it, its controlling effect over the main thermostatic control will be limited to 50 per cent. Higher resistances can be utilized to increase this ratio.

This arrangement is frequently employed in air-conditioning equipment, where it is used for low-limit control. In a hot-air heating system, for example, if the valve on the normal heating supply is closed off completely when the thermostat is attempting to bring the temperature of a room down, the fan would still continue running and blow a supply of cold fresh air into the occupational zones. Whilst this would lower the temperature, the rapidity with which the cold air did this might cause discomfort to the occupants. A low-limit thermostat could be included and set to a temperature slightly above that causing discomfort so that when the main controller shuts off, the low-limit thermostat takes over, reopening the valve or controller before the low-temperature conditions became uncomfortable. Similarly, this system can be used as a high-temperature limit controller to ensure that at no time is the incoming air uncomfortably hot.

As both high- and low-limit switches are set for extreme temperatures, under normal conditions they resume the fully satisfied station and do not bias the basic control circuit. Most manufacturers can supply a range of thermostats to go with this type of electric proportional control system. Normally they are vapour-tension devices, either of the self-contained variety for room control, or remotely operated by phial and capillary suitable for ducting and water-temperature control. The system is equally suitable for the control of pressure and humidity, so that the same proportional control circuit can operate where all three of these conditions have to be regulated, as, for instance, a full air-conditioning and heating system.

Electronic systems

A natural development of the electric method of proportional control described in the last section is electronic proportional control. This is also based on the Wheatstone bridge circuit. The general problem of electronics is beyond the scope of this book, but the following brief description of the electronic proportional control will indicate its principle of operation.

The circuit of a typical system is shown in Fig. 8.4. The temperature element is an electrical resistance thermal unit (T1) which is connected in one arm of the bridge. The out-of-balance current across the bridge that results from a change in the temperature of the thermal element provides a 'signal' which is amplified and then passed to the phase detector. This detector decides whether the signal has been caused by a rise or fall in temperature, and then energizes the corresponding relay (R1 or R2) in order to drive the reversing motor in the direction required to correct the change. As in the electric system, the motor, in addition to driving a valve or damper unit, operates a feedback potentiometer (RB) which forms part of the bridge network. The motor continues to run until the wiper of the feedback potentiometer has moved by an amount equal to the deviation created by the temperature change at the thermostat. The 'error signal' in the bridge circuit is thus cancelled out and the bridge is re-balanced and the motor stopped. The bridge can be pre-set for a particular installation by adjusting RD and RP.

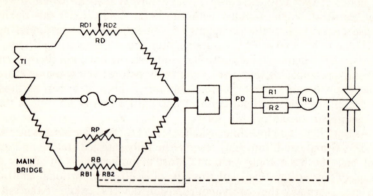

Fig. 8.4. *Electronic proportional control system using an electrical resistance thermal element (T1).*

T1 *Temperature element*	**PD** *Phase-sensitive detector*
RD *Desired value potentiometer*	**R1** ⎱ *Output relays*
RP *Proportional band potentiometer*	**R2** ⎰
RB *Rebalancing potentiometer*	**RU** *Actuator*
A *Amplifier*	

This type of control arrangement offers some advantages. The thermostatic controller has no moving parts and a very low thermal mass, and so can very quickly detect slight changes in temperature: electronic systems tend to be extremely sensitive. They also lend themselves to very complex control problems where many factors can be taken into account all of which can be fed into a central control system where a variety of adjustments and modifications of control conditions can be effected.

There are many variations of this basic principle, each of which has its particular advantage. One example incorporates integral action as well as proportional action. In this type of control, a change in temperature sets in motion the normal proportional action in which the motor operating the valve or damper unit is moved proportionally to the change of temperature sensed at the temperature detector; simultaneously, an integral action is derived from a timing switch which causes the movement of the valve to continue at a selected speed until the room temperature has been restored to the exact desired value.

Pneumatic systems

Air-operated systems are commonly used in large heating and process schemes. They can be used to control not only temperature, but also other variables such as humidity and pressure. We shall here restrict ourselves to their application to thermostatic control. They have several distinct advantages. All risk of fire due to the control itself is eliminated, even when combustible or highly inflammable liquids are being controlled, as for instance in oil refineries. The simplicity of the control components makes the system adaptable to many control

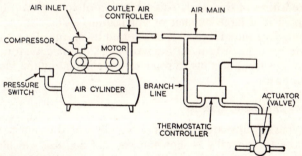

Fig. 8.5. Basic requirements for air supply on a pneumatic control circuit.

sequences and a complex system can be built up at relatively low cost. In recent years it has found some success in the control of domestic heating systems in America where advance control techniques using zoning schemes and day and night temperature changes are in frequent use.

Fig. 8.5 shows the basic elements of a simple pneumatic system. It comprises a small compressor capable of producing air at pressures up to 1 bar. In addition to the normal requirements of such a compressor, it is essential that a good filter be fitted, together with a pressure regulator and, in some cases, a means of de-humidifying or drying the air. The compressor unit compresses and stores the air in the air cylinder until it is required for use by one of the thermostatic controllers in the circuit. The air is fed to these controllers by the main pipe line, which is generally made of copper. For each control circuit a line is taken from the main supply to the thermostatic controllers; these change the pressure in accordance with the movement of the thermal element and supply air at a reduced pressure to an actuator, i.e. a valve or pneumatic motor, causing it to open or close as circumstances require. In the normal arrangement the controlled actuator assumes a position proportional to the pressure supplied to it by the thermostat. In this way the pneumatic system as a whole will operate as a proportional controller. If the valve is bellows-operated, as shown in Fig. 8.6, its position will be determined by the pressure from the controller, and the forces opposing this pressure. The latter are made up of the return spring on the bellows, the extraneous forces such as the opposing action of the valve and the liquid flowing through it or, in the case of a motor-operated damper, the friction of the damper mechanism itself, and any opposition set up by draught or wind.

In the simplest form of pneumatic-control system the actuator does not always assume a positive position when working under a variable load: it assumes a position in accordance with the direction and the total opposition of this load. This limitation of control can, however, be accepted in a large number of applications. Where more positive positioning is required a relay is introduced by means of which greater forces are brought to bear on the movement of the actuator, making this more positive and thus to some extent eliminating the limitations just mentioned.

Valves for pneumatic systems

Two types of valve are employed, normally closed or normally open ones. These are generally as shown in Figs. 8.6 and 8.7, which are self-explanatory. When a large differential pressure is to be controlled and when this pressure would be sufficient to exert considerable influence on the bellows, a double-beat valve, Fig. 8.7, is employed. In this the differential pressure is to a large extent balanced out. The pressure on the inlet side exercises a force upwards on the upper valve and downwards on the lower valve; thus the two valves are virtually in equilibrium and the pressure to open or close the valve required from the pneumatic-ally operated bellows is comparatively light.

Damper units

These operate on a similar principle to the pneumatically operated valve but are designed to act by the motion of a lever which can be coupled to the damper unit, or to any other device where positive angular displacement is required in relation to a change in temperature.

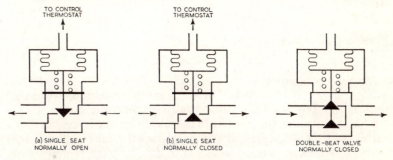

Figs. 8.6 (left) and 8.7 (far right). Three basic valve arrangements for use in a pneumatic control circuit.

Controllers

A simple type of thermostatic controller for use with pneumatic systems is shown in Fig. 8.8. The air supply from the compressor is split so that one leg goes to the control valve, the other terminating at a small nozzle in the thermostat which normally vents the air to atmo-sphere. When flow through the nozzle is unrestricted, the air pressure in the branch line will not build up and the control valve will remain in its depressurized position, i.e. fully open or fully closed, according to type.

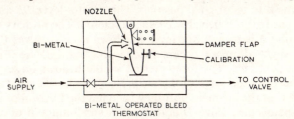

Fig. 8.8. Bi-metal operated room thermostat for use with a pneumatic control circuit.

If a small flap is placed near the nozzle so that it restricts the flow through the nozzle, a pressure will build up in the air line until, if the nozzle is completely closed, the air pressure in the line equals the mains pressure. The control valve or actuator gradually reacts to the rising pressure, assuming a position between its fully closed and fully open positions; finally, when the flap completely seals the nozzle, the control valve assumes one of its extreme positions. The damper flap is operated by a bi-metal strip which is first set by the adjustment provided— generally a range knob scaled in temperature. From then onwards the movement of the bi-metal in response to a change in temperature operates the valve controlling the heat supply in the manner already explained.

Non-bleed controller

The type of controller just described is called a *bleed* controller, because the air is allowed to 'bleed' or escape to atmosphere at all times other than when the flap completely closes the nozzle. In the non-bleed controller, the pressure in the branch air line is regulated by a system of valves which prevent all flow through the line except when the branch-line pressure is being increased. A diagrammatic representation of this type of controller is shown in Figs. 8.9, 8.10 and 8.11. In Fig. 8.9 the conditions are those for a satisfied control system: the supply valve *C* and the exhaust valve *D* are closed and the pressure in the branch line is steady. In Fig. 8.10 a slight temperature increase in the controlled condition causes the bellows *A* to expand. Since the bellows is fastened rigidly at the top, it expands downwards, rotating the lever *BE*: exhaust valve *D* now becomes the fulcrum of lever *BE* and valve *C* is lifted. A small amount of air enters from the main, increasing the pressure in the branch line and pressure chamber. The increased pressure acts on the diaphragm *G* to oppose the force of the thermal bellows *A* and returns it to its original position. This rotates lever *BE* and closes port *C*. During this time valve *D* remains closed. The primary function of the spring *F* is to hold the lever *BE* down against valves *C* and *D*. When a slight decrease occurs in the temperature (see Fig. 8.11) bellows *A* contracts. The force exerted by the bellows on lever *BE* is overbalanced by the force of the air pressure under the diaphragm *G*. The valve *C* at the air-main port now becomes the fulcrum of the lever *BE* and the exhaust valve *D* is opened. A small amount of air is exhausted from the system so that the branch line and chamber pressure is decreased. This reduces the upward force on the diaphragm *G* and allows the bellows to expand until the exhaust port is closed.

Whenever the branch line reaches the pressure called for by the thermostat, both valves are closed, the system is in balance and no air is used.

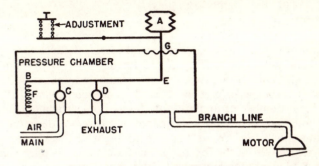

Fig. 8.9. Non-bleed controller in satisfied condition. (Honeywell Controls Ltd.)

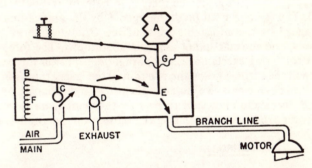

Fig. 8.10. Non-bleed controller—increase in measured condition. (Honeywell Controls Ltd.)

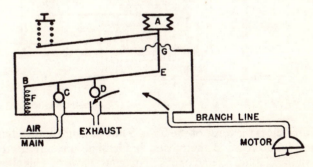

Fig. 8.11. Non-bleed controller—decrease in measured condition. (Honeywell Controls Ltd.)

Relay system

As earlier stated, the motor or pneumatically operated valve changes position owing to a balancing of several forces. Because some of these forces, e.g. those due to friction in a linkage system, are undefinable, the actuator cannot be guaranteed to assume any one position in relation to a change in pressure caused by the controlling thermostat. In a large number of applications this is not particularly important but where exact positioning is required or where it is desirable that the actuator moves quickly in response to slight changes in pressure, a servo-mechanism in the form of a pneumatic relay is introduced.

Fig. 8.12 shows the Honeywell Gradutrol system which is typical of this arrangement. Valve *B* feeds air to the motor when the thermostat branch-line pressure increases and valve *C* exhausts air from the motor when this pressure decreases. When the motor is in the position required by the thermostat the relay is in balance and both valves are closed. The force exerted on diaphragm *A* by the branch-line pressure is balanced by the adjustable force of the Gradutrol spring and cup assembly, and the start-point adjustment spring *G*. The force that the Gradutrol spring exerts on the diaphragm depends on the position of the cup in the spring and on the position of the lever arm *E*. An increase in the branch-line pressure increases the force on the diaphragm. Supply valve *B* opens and allows air to pass from the air main to the motor. The motor arm moves, increasing the tension on the Gradutrol spring

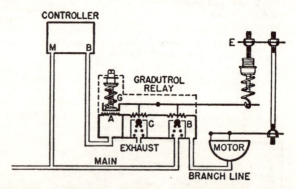

Fig. 8.12. A simple Gradutrol system. (Honeywell Controls Ltd.)

until it balances the force of the branch-line pressure on the diaphragm. Similarly, a decrease in thermostat branch-line pressure decreases the force on the diaphragm and opens the exhaust valve *C*. This allows air to exhaust from the motor until the force of the Gradutrol spring decreases enough to balance the reduced force on the diaphragm. A very small change in branch-line pressure can produce any required pressure on the motor between 0 and 1 bar.

By use of this relay the actuator is capable of producing full power in order to assume any position called for by the thermostat controller, and is thus able to overcome any friction loads caused by valve-stem packing or mechanical linkage friction. As soon as the thermostat controller changes its pressure, its signal to the motor, no matter how weak in terms of pressure change, is converted to high pressure, i.e. anything between 0 and 1 bar, which moves the motor positively until, by the linkage between the motor and the relay valve, a balance of pressure is restored within the relay. This relay system is also useful, and in fact necessary, where it is required that two valves working in the same system should never be open at the same time. The linkage between the pneumatically operated valve and the relay gives a positive control over the valve or motor position.

In addition to the relay described above, there are other versions of this mechanism capable of converting any pneumatic system from straight proportioning control to any of the other types of control, i.e. (1) on-off, (2) proportional, (3) proportional plus reset or integral, (4) proportional plus reset plus derivative control.

(1) and (2) are to be found on industrial and commercial heating and ventilating processes, furnace controls and other applications, whilst (3) and (4) are normally confined to industrial process systems only.

9

BASIC HEATING CONTROL ARRANGEMENTS

EVERY complete heating circuit comprises the following: fuel, a means of converting the fuel into heat, an appliance or heat exchanger into which heat is conveyed, and a means of distributing the heat thus produced to the parts of the building or process where it is required. At each stage of this system adequate control is essential if it is to operate safely and economically and, at the same time, provide the continuous flow of heat needed in relation to the varying demands. The controls employed for obtaining this automatic operation vary from one fuel to another and depend also on the type of heat exchanger or appliance in use and the special requirements of the type of heat

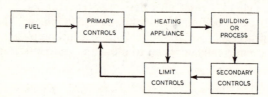

Fig. 9.1. The basic control loop in thermostatic heating control.

being provided. Nevertheless, certain basic features remain the same in each system, and these are shown diagrammatically in Fig. 9.1.

Fuel is first made available to a burner; this may be gas fed to a gas burner, solid fuel supplied to an open or chain-grate stoker, or oil fed into an oil-pressure jet burner. The same features also apply, to a limited extent, to electric energy made available to heating elements. In every case there is a need to provide automatic control to the flow of fuel so that whatever the heat demand the fuel is burned or disposed of safely and economically. These controls are generally known as 'primary controls'.

Primary controls

These must measure-in the fuel so that it can be disposed of economically by the burner or other means of heat conversion, must regulate the combustion process and, if necessary, stop the flow of fuel if the conversion to heat is not proceeding satisfactorily.

High-limit controls

The function of these devices is to protect the heat exchanger or other component accepting heat from the burner or heat source, and to ensure that the temperatures at this stage do not rise above a pre-determined safe maximum. In some instances these controls are not set to the maximum temperature to which the appliance can work, but to a convenient temperature above which heat becomes uneconomical, or unnecessary. Whilst for the most part high-limit switches will be thermal devices they can include pressure-operated valves, switches, and level controls. These latter items are essentially for use on steam-raising plant, where both correct water level and pressure are of importance in the safe operation of the system as a whole.

Normally the thermal devices used as high-limit controls can take the form of standard thermostats.

As these controls have to sense conditions within a heat exchanger compartment they are invariably of either the remote-control type or the duct-mounting type. The remote-control variety use vapour-pressure or liquid-expansion systems with separate sensing head and capillary, operating the control switch through the medium of a bellows or diaphragm. Duct-mounting thermostats are either of the coiled bi-metal or the rod-and-tube thermal variety. Being high-limit switches they are concerned only with an accurate cut-out temperature, and differentials are normally set wide so that there is no possibility of rapid re-cycling as the thermostat cuts in and out if a high temperature condition persists.

A purely high-limit thermostat is required to work only in abnormal conditions. Therefore, once a high temperature has been reached it is important that the burner or other combustion process is switched off until the temperature has fallen to the required level. It is for this reason that many high-limit thermostats have a permanent lock-off position, so that once they have operated they remain off and thus indicate that an abnormal condition exists. The manual-reset variety are either modified versions of the standard thermostat, incorporating a trip action so that once switched off the thermal element is prevented from remaking the circuit, or of the fusible-link type. In the latter, instead of a standard thermal system the sensing device takes the form of a solid material which becomes liquid when a certain temperature has been reached, and in doing so allows a spring action to come into operation and move the contacts to the open position. On a drop in temperature the material resolidifies and the spring-operated contacts are reclosed by hand.

In purely high-limit thermostats it is an advantage to have a *fail-safe* type of thermal system, i.e. if the thermal mechanism fails the switches automatically move to the off position. Failure to the on position would render the high-limit thermostat inoperative, a fault which might well remain undetected until a dangerous condition had developed.

High-limit controls are always installed in the control circuit in such a way that they over-ride all other controls in the heating cycle (Fig. 9.2). In this way the heat exchanger or other appliance can only supply heat against the demand from other control systems as long as safe conditions prevail.

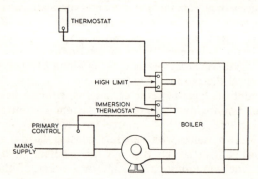

Fig. 9.2. Simple control scheme for an oil-fired boiler.

Heat output controls

These are also thermostats situated on the heating appliance and controlling its output. Unlike the high-limit controls, they are not necessarily set to the maximum temperature the appliance is capable of providing, but can be varied in setting to meet varying demands on the appliance. In a warm-air furnace or heat exchanger supplying hot air to a ducted system, these thermostats can be varied to provide a hot-air temperature in relation to the heating demands, and in the hot-water system they can be set to meet the demands for hot water whether this be for domestic use or central heating. Their operating differential is of importance, and in many cases can be adjusted for the particular site conditions. Adjustable range is also necessary so that this can be set for the demands of the moment.

In some applications where the appliance control is sited in the heat exchanger it can fulfil the dual function of appliance and high-limit control.

The decision to use two separate devices or one device combining the two functions really depends on the conditions likely to be encountered. Where the possibility of danger is very small, a dual purpose device is acceptable, but where an uncontrolled heat input may be hazardous, a separate high-limit device with fail-safe features is essential.

Low-limit controls

These devices are not so frequently used as high-limit controls, although their function is similar in that they sense conditions within the heat exchanger and on a fall in temperature below a predetermined limit will switch on the heat source regardless of the demands, or lack of demands, being made by the secondary controls. They are of particular use in hot-air systems where air is continuously supplied to a building as a means of ventilation or heating. Where the secondary controls are satisfied and are not calling for any heat the continued supply of forced air to the building would cause the temperature to drop below comfort condition unless some heat was put into it. The low-limit thermostat monitors this; if the secondary controls are in the off position and the air entering the building falls below a predetermined temperature, the low-limit thermostat will switch on the heat source bringing the air to minimum comfort temperature. Similarly, the low-limit thermostat can be used on hot-water circuits when the secondary controls have been switched to the off position during the night or weekend, but where frost may have reduced the water temperature in the heating circuit to near freezing. A low-limit thermostat would sense this condition and cause the heat source to be switched on even though the secondary controls were not demanding heat.

Secondary controls

In domestic heating the unit heater, such as the gas fire convector heater, electric convector heater, or solid-fuel heating appliance is virtually self contained, and any thermostatic control fitted to it can be an integral part of the appliance. In these instances, the thermostatic control is a simple one in which a temperature-sensing phial is arranged to have an immediate reaction on the source of fuel, or the control of its burning. Satisfactory control can be provided so that the room in which the appliance is sited is held at any selected temperature.

Any heating system larger than this normally entails a heat exchanger capable of producing sufficient heat to be distributed to a number of rooms or over a large area, and once heating problems of this size are encountered it invariably follows that thermostatic control becomes a more complex process. As soon as heat is carried away from the appliance to heat remote rooms or sections of a building, thermostatic control has to be effective at these distances from the appliance and must be able to transmit signals back to the heat source, controlling it in relation to the conditions at these remote points. The secondary controls are additional to the primary and high-limit controls which, as we have seen, ensure the safe and satisfactory operation of the appliance itself.

In its simplest form secondary control can take the form of an on-off thermostat sited at a convenient point and transmitting an on-off action to the heating appliance. Generally as the size of the appliance and the heating problem increases so the control problem becomes more complex, calling for accurate control of several zones, each zone having its peculiar requirements.

Consider first the simplest form using an on-off thermostat. Its effectiveness will depend on the differential of the thermostat and the cumulative effect of the various lags that can take place in the heating system. The use of accelerator-type thermostats will provide some improvement as already explained in Chapter 7.

On larger schemes the secondary control will take the form of proportional or integral systems where the heat input is proportional to the deviation in the controlled temperature. In central-heating systems inside-outside temperature control can be used where the flow temperature in the hot-water circuit is proportional to the outside temperature, and in even more complex systems these controls can be applied to several zones, each of them obtaining their heat from the one source.

The same considerations can equally apply to process-control requirements where one heating source is expected to supply the varying heat demands for several separate processes. Thermostats can be combined with timers and other regulating devices, all of which make up the secondary control circuit capable of drawing from the one heat source the input required to maintain control conditions in their particular zone. It is these combinations that make the most interesting and most useful contribution to the automatic-control systems described in the following pages.

DOMESTIC HEATING
CONTROLS

THE increasing emphasis on home comfort in terms of warm houses has
been a feature of the higher standards of living achieved in the last
decade: the warm house is now regarded more as a necessity than, as it
was in the past, a luxury. This demand has been created at the same time
as fuel costs generally have risen, so that it has been accompanied by
increasing attention to heat conservation, both by good thermal insulation of buildings and also by the economy of fuel consumption that
results from thermostatic control of the heating appliance. Most types
of heating appliances, irrespective of the type of fuel they consume,
now either have a thermostat as a standard fitting or can be obtainable in
alternative forms so equipped.

For heating existing dwellings not designed and equipped with
modern heating installations, most fuels can be utilized in a range of
portable appliances. Similarly, the advent of small-bore central heating
has made this type of house warming available for easy installation.
New houses, whether individually and architecturally designed, or part
of estate development, usually have central heating in one form or
another. As a result of this trend large numbers of specialized thermostatic devices have been developed; and certain basic trends towards
well-defined distinct classes have become apparent during the last
decade.

PORTABLE APPLIANCES

In this class, the predominant types are free-standing oil-burning
heaters and the electric heater either of the convection or radiant-
element type. The free-standing oil heater offers little opportunity for
full thermostatic control. Being portable, it incorporates its oil supply
in the form of a barometric feed, or fuel tank from which oil is drawn by
capillary attraction through a wick or by constant level feed. In both
instances a fuel supply equivalent to some 24 hours running is generally
supplied, with a manually adjustable wick or flow-rate adjustment to
control the heater output. The technical difficulties of supplying thermostatic control have so far precluded its introduction in the majority of
appliances available.

Electric portable heaters

The electric portable heater is by far the most popular of the portable heaters available, and there must be very few homes left that do not possess at least one of them. Their popularity is undoubtedly due to their low initial cost in relation to their heat output and, as far as the radiant versions are concerned, their ability to supply heat quickly to a small part of a room where the heating of the whole room is either uneconomic or unnecessary. In this respect the type resembles the traditional coal fire which brings a small area in its vicinity up to comfort conditions without heating all the air in the room to a similar standard. In recent years, considerable attention has been given to the development of purely convector heaters which are designed for more continuous use and for heating the entire room, rather than a localized part of it as the radiant heater does.

Radiant heaters

Two types of radiant heater are available, generally referred to as fire-bar and reflector fires. The fire-bar consists of an element wound into a continuous spiral and laid into a fire clay slab. This type of fire produces some 50 per cent of its energy in the form of radiant heat, the other 50 per cent being given off as convection heat from its ability to warm the air that passes over the high-temperature surfaces of the fire clay and the surrounding metal supporting members.

The reflector fire can have one or two elements wound round a fire clay support or can take the form of a heater element both of which achieve high temperature and glow to a cherry red. These elements are fitted in front of a highly polished reflector shaped in the form of a bowl or trough which concentrates their radiant heat into a beam. These heaters, having a low thermal mass, are capable of producing up to 70 per cent of their output in the form of radiant heat.

Thermostatic control of this type of fire presents some difficulty, not only because, as stated earlier, they do not as a rule heat all the air in the room, but also because most of their heat is emitted as radiant heat. The normal room thermostat is designed to operate on air temperature; therefore merely siting a thermostat on the appliance would achieve little effect. Similarly, a thermostat placed remotely from the fire and in line with the radiant heat would operate at a false temperature and switch off the fire leaving the room largely unheated.

Sometimes, however, these fires are used for full time heating of a room and then some thermostatic control can be effected by siting a thermostat out of range of the radiant heat from the fire and working purely on air temperature. Several types of thermostat are available for use with portable electric heaters in these conditions. They are specially designed so that they can be plugged into the power-output socket and the fire plugged into another socket built into the thermostat. This places the thermostat in series with the heater and effects on-off control. In such applications a degree of relative control is inevitable because:

(1) The thermostat will be sensing temperatures in the vicinity of the wainscoting of the room, and these are normally lower than the rest of the area. This difficulty has been overcome in some designs in which the thermostat is mounted higher up on the wall.
(2) Where a fire may be providing anything up to 70 per cent radiation, the air temperature does not really give a true picture of the comfort provided.

With a purely radiant heater there is the disadvantage that as soon as the thermostat switches off the heater there is an immediate loss of radiant heat which has a bad psychological effect on the people within its range. It is also found that before such a fire can be switched off completely without discomfort, the air in the room must be heated to a temperature in excess of that which would normally be necessary with purely convection heating.

Convector heaters

Electric convector heaters lend themselves more easily to thermostatic control and, in fact, most of them have thermostats either as a standard fitting or as an optional extra (Fig. 10.1). These heaters are offered in a variety of designs and shapes, but consist essentially of a metal cabinet having a cold-air inlet at the bottom, and a warm air outlet at the top. Between these two openings is sited a heating element and a series of baffles arranged so that maximum air entrainment is achieved by passing the air across the heating element on its way through the cabinet. Whilst the majority of these convectors rely on the natural draught created by the heating to circulate the air through the heater and then into the room, some of the larger models are supplied with fans to assist this action. The essence of good design in this type of appliance lies in ensuring that the amount of air being passed through the heater is sufficient to keep the casing temperature down to a safe figure and, at the same time, to ensure a rapid supply of heated air to the room. Generally the output of these heaters is in the order of 95 per cent convection heat and 5 per cent radiated heat from the surface temperature of the cabinet.

Thermostatic control of this type of appliance can be achieved satis-

116

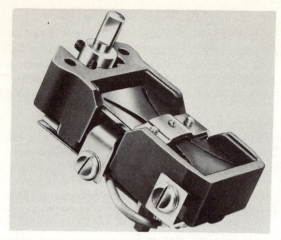

Fig. 10.1. Convector thermostat. (Robert Maclaren & Co., Ltd.)

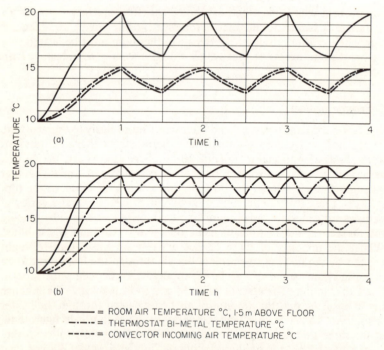

——— = ROOM AIR TEMPERATURE °C, 1·5 m ABOVE FLOOR
—·—·— = THERMOSTAT BI-METAL TEMPERATURE °C
- - - - = CONVECTOR INCOMING AIR TEMPERATURE °C

Fig. 10.2. Effect of thermostat differential and local heat. (a) No local heating, thermostat differential 2·2°C, thermostat setting 16°C. (b) 4·4°C local heating, thermostat differential 20°C, thermostat setting 20°C.

factorily provided that one or two basic factors are borne in mind. The problem in heating a room with natural convection heat is that some stratification of room temperature takes place and, therefore, the thermostat fitted in the appliance, probably only a few centimetres above floor level, will not be sensing room conditions as experienced by the people in the room who are mainly sensitive to temperatures some 1·5 m above floor level. Some indication of this relative control is shown in Fig. 10.2(a). The thermostat used had a differential of 2·2°C and was set to cut out at 15·5°C and cut in at 13·3°C. It will be seen that the effect of this setting and differential was to produce a room temperature which fluctuated between 21°C and 17°C. In this type of application with the thermostat completely protected from any extraneous heat, the cycling is very slow because of the long time taken for the thermal element to heat and cool, which, in turn, is due to the very slow changes in temperature of the incoming air at floor level.

Accelerator thermostats

To overcome this basic problem, the manufacturers of these appliances have adopted some form of accelerator thermostat (generally of the bi-metal type), the basic principles of which have already been discussed in the chapter on on-off thermostats. This principle employs the application of a second source of heat to the thermostatic element each time it calls for heat. Such a source of heat can be found by: (1) passing the electrical load of the convector heater through the bi-metal and thus causing a rise in temperature; (2) incorporating a small heater in close proximity to the bi-metal; or (3) deflecting some of the heat from the main heater element of the convector heater to the thermostatic metal.

Each of these methods has its own advantages and disadvantages; in the main, the last appears to be most widely used. In this system, every time the thermostat calls for heat a regulated temperature increase of the bi-metal itself is caused by back radiation or conducted heat from the heating element. Thus, the thermostat does not have to wait for a rise in temperature of the incoming air sufficient to cause it to break circuit, but breaks circuit as soon as the additional heat from the extraneous source causes the temperature of the bi-metal to rise. This produces an effect as shown in Fig. 10.2(b). Here it will be seen that the local heat causes the thermostat to operate at a temperature closer to the room temperature at occupation level, rather than to that of the incoming air. It will also be observed that the introduction of local heat causes the thermostat to operate in a series of short cycles, thereby maintaining room temperature at a much steadier level.

Several permutations of thermostat differential and local heat are employed, according to the construction of the appliance, the degree of control expected from it, and the final cost.

Fan-assisted heaters

Fan-assisted electric heaters use similar bi-metallic devices to control their output. The thermostat generally works on a fine differential of 1° or 2°C, and senses incoming air temperature. With the more rapid movement of air through the heater caused by the fan this temperature is more representative of the average room temperature than it is with a free-convector heater. In addition to controlling the heating element, the thermostats employed in fan heaters are sometimes arranged to operate changeover switches so that when the room thermostat reaches the desired level the heater is not cut out completely, but is switched to a lower fan speed and a lower-wattage heat output.

With this arrangement one is less conscious of the thermostatic action of the heater, and a better room control is provided. As with the natural convection heater, most fan units provide a further thermostatic device in the form of a safety cut-out. This is pre-set to trip out the heater and fan completely if an unusual rise in temperature is experienced within the appliance.

Oil-filled radiators

A third type of portable electric heater takes the form of an oil-filled radiator. This type is widely used for continuous room heating, and as a background heater to supplement other forms of heating. Basically it consists of a steel radiator filled with oil which obtains its heat from a thermostatically controlled immersion element in the base of the radiator. A thermostatic switch is housed on the side of the unit suitably protected in a vented compartment through which air from the room is drawn. Being sited low in the room, this thermostat is sometimes off-set to provide relative control of air temperature. In addition, a safety cut-out is sometimes provided sensing the temperature of the oil in the radiator and locking out if an abnormal rise occurs. This is generally of the manual reset type so that once locked out it will remain so until the cause of the over-heating has been detected and the safety switch reset.

Storage heaters

The most recent addition to the range of domestic electric-heating appliances is the storage heater; whilst its weight really excludes it from the portable class, the fact that it is not normally fixed in its position allows it to be described in this general group. The appliance consists essentially of electric heating elements embedded in a block of specially prepared concrete or other material having a high thermal density. This mass is heated to a very high temperature, and then discharges its heat slowly over a period of hours. This type of heater has been specially developed to take advantage of the low electricity tariffs applicable in most areas through the hours of darkness. This allows the heater to be charged with heat during the night and switched off in the morning, after which it gradually discharges the stored heat during the daytime.

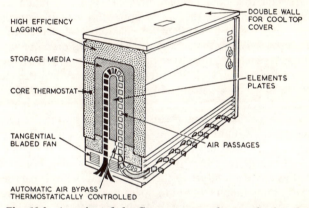

Fig. 10.3. A section of the Constor storage heater, showing the principle of operation and the controls.

In its simplest form the heater consists of the thermal block, as previously described, suitably protected by a metal case which is louvred so that a natural convection current will pass through the casing taking heat from the block and passing it out into the room. In this type of heater the thermostatic control is limited to an excess temperature device which comes into force if the storage block reaches an abnormal temperature. A second thermostat can be fitted which limits the temperature of the block so that the user can limit the amount of heat taken into storage during the night, and discharged during the following day.

A more complex form of storage heater is constructed so that its heat output is much more flexible. Of this type, the Constor (Fig. 10.3) represents one of the more advanced patterns. It will be seen that the storage unit is heavily insulated so that the external surfaces remain comparatively cool. Air is entrained through the base of the heater by fan and is drawn across the storage medium and discharged through louvres in the base of the appliance. The heater is charged with heat during the night in the normal manner, but if there is no demand for heat during the day the leakage of heat is considerably reduced by the heavy lagging. With the thermostat demanding heat and with the fans switched to full speed, the heater is capable of raising room temperature in a few minutes. From then onwards the fans are controlled by the room thermostat and heat is drawn off from the storage unit as it is required.

The so-called core thermostat used in storage heaters in fact works on a relative temperature rather than the actual core temperature. This is due to the fact that core temperatures can reach as much as 800°C, and normal commercial thermostats are not capable of providing effective control at this temperature. Therefore, a degree of relative control is used in which the core thermostat works on a known relative condition between the temperature of the insulation and the temperature of the core. Adjustment of this thermostat can control the maximum temperature to which the storage block rises, and thus limit the amount of heat absorbed during the night. In this way some anticipation for a mild day with less heat demand can be made by setting the thermostat down the previous night.

A further thermostat device is incorporated in the air channel before the final discharge into the room. This consists of a bi-metallic coil which senses the temperature of the air as it is about to leave the heater. This device limits the maximum temperature of the air discharge by, if necessary, mixing the air passing through the core with some air straight from the room. In this way, the air discharge temperature is kept at a comfortable maximum.

The room thermostat for controlling the output of the heater during the day can take the form of a standard air thermostat. For the type of heater where the thermostat merely switches the fans on and off and does not control the initial heat input into the heater a standard thermostat with accelerator (see Chapter 7) can be used with advantage. This especially applies when the storage heater is fitted in a confined space and is capable of raising room temperatures rapidly. However, where a room thermostat is wired in such a way that it can control the heat input into the storage heater during the heating up cycle, an accelerator thermostat should not be used, as it could limit the amount of heat stored during the off-peak period to an amount insufficient for the next day's requirements.

Gas space heaters

The recent development of the combined radiant and convection heater fired by town gas has given a considerable stimulus to gas heating.

The increased efficiency obtained by utilizing the thermal mass of the fire to provide convection as well as radiant heating has made this type of appliance very economical, and some manufacturers state that the running cost can be comparable with that of a solid-fuel fire.

As in the case of the electric radiant heater, the earlier gas fire did not lend itself to thermostatic control. In the type with appreciable convection heating, however, thermostatic control can be provided as an integral part of the appliance. The general design of these appliances is shown in Fig. 10.4. Radiant heat is provided by the gas-heated ceramic material which quickly becomes incandescent. Convection heat is obtained by passing air round the back of the combustion zone and the

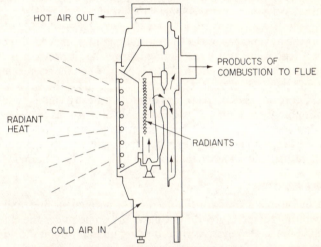

HOT AIR OUT ◄——

PRODUCTS OF
COMBUSTION TO FLUE

RADIANT
HEAT

RADIANTS

COLD AIR IN

Fig. 10.4. Modern gas-fire chassis showing heat exchanger sited above radiant-heat zone. (Flavel.)

heated ceramic block and into a heat exchanger before passing into the room. It is heated by means of finned members and other aids to maximum heat transfer. The products of combustion are separately flued and pass away to the chimney.

It is a convenient fact that the air is drawn in from the room at the bottom of the appliance in close proximity to the gas supply and controls, as this makes it comparatively simple to install a thermostatic valve in the gas supply and to have a phial from this valve sensing the temperature of the air being drawn into the convection channels. The air, having been drawn from the room, is representative of room temperature and is therefore an effective guide as to whether comfort conditions have yet been reached.

Some fires incorporate a thermostatic valve whose thermal element is a vapour-pressure system, consisting of a bellows within the valve area connected to capillary tubing which is sited in the airstream and acts as a sensing phial. A temperature-selection device is set by a range knob, the rotation of which transmits compression to a spring directly in opposition to the bellows. By this arrangement the thermostat can be arranged to provide a valve-closing action at any selected temperature between 16° and 32°C.

Owing to the compact nature of the appliance it is sometimes difficult to locate the phial position so that the sensing portion only receives the true temperature of the incoming air, and in most cases it has to tolerate some back radiation from the heated parts of the appliance. To offset this characteristic it is therefore sometimes necessary to allow the thermal system to operate some 11°C in excess of real room conditions.

One of the advantages of the thermostatic control provided on these appliances is that, being a direct-acting thermostat, it provides a degree of modulation rather than on-off control. Thus, when the room temperature attains the desired value the thermostat reduces the gas rate to a nominal figure so that complete loss of radiant heat is not experienced. This undoubtedly has a good psychological value for, as stated earlier, the sudden loss of all radiant heat has a depressing effect on people close to the fire, who generally react by turning the fire up to produce an air temperature much higher than would normally be required.

Solid fuel appliances

Solid fuel appliances, with the need for a permanent flue, and their high operating temperature, do not normally come under the category of portable units. However, as free standing units for domestic application they are widely used in front of existing fire grates. Their design has been subjected to a series of improvements over the last decade, with the result that there is a choice of highly efficient units now available. Several types of fully enclosed fires have been fitted with thermostatic control of one type or another in spite of technical difficulties.

The rate of burning in the enclosable solid-fuel appliance depends on the rate at which air is allowed to enter the combustion zone; therefore any thermostatic control must be able to vary the air input to the fire in relation to some temperature condition. The heat output from the heater is obtained as: (1) radiant heat from the incandescent fuel through the doors of the appliance; (2) radiant heat from the metal castings of which the heater is constructed; and (3) convected heat obtained by the air passing across the heated surfaces. Therefore, any thermostatic control attempted must take into account these three sources of heat.

One difficulty lies in the fact that there can be a considerable time lag from the moment that the thermostat senses a drop in room temperature and opens the air to the fire, to the time when the fire recovers from its damped down rate and achieves a high incandescence with its maximum output of heat. There is a further problem in that when the room is up to temperature and the fire is filled with incandescent fuel any throttling of the air supply to the fire will not have an immediate effect on its output. In fact, the incandescent fuel will continue to discharge heat for a long period which will necessarily cause further rises in temperature in the room.

Thermostatic control of these appliances therefore takes a relative form and in some instances is arranged to provide a controlled rate of combustion rather than exact control of room temperature. As was stated earlier, control is achieved by varying the air input into the fire, and it is therefore necessary to seal off all possible air-entry points other than through controlled air intake. The control generally takes the form of a disc acting as a flap which can completely close the air intake. The flap is mounted on a lever which, in turn, is tilted by some form of thermal device. The prime mover in the thermostat (sensing a mixture of incoming air temperature and casting temperature) is a gas-charged system similar to that described for the domestic hot-water boiler.

DOMESTIC CENTRAL HEATING BY HOT-WATER CIRCULATION

The distribution of heat to all parts of a house through the medium of hot water circulating through radiators is still by far the most widely used method of providing central heating, and recent advances in design have increased the usefulness of this system making it suitable not only for new premises but for easy installation in existing ones.

The use of thermostatic control both for the boiler and also for the control of the heat distribution is considered essential, and does in fact contribute a considerable amount towards the increased efficiency of these systems.

Until the advent of the small-bore circuit, domestic central-heating systems depended on a heating appliance, generally of the solid-fuel type, feeding radiators in various rooms through large-diameter pipes. The piping was arranged to provide a thermo-syphon system whereby the heated water circulated throughout the house by gravity. The thermo-syphon action is a comparatively weak pressure difference between the rising hot water and falling cold water, so that the pipework had to be designed for as low a resistance to flow as possible.

This necessitated pipes of a diameter of about 38 mm, with easy bends and gradual falls. Such pipework tended to be unsightly by modern standards when installed in a new house, and was even more difficult to install in an existing house. However, large numbers of these systems are still in existence working comparatively efficiently and some thermostatic control is generally effected at the boiler.

The type of control mostly in use was a simple device consisting of a bellows filled with wax and sited in the waterway of the heat exchanger in such a manner that changes in water temperature effected an expansion or contraction of the wax. This caused the bellows to change length and therefore to open or close the air flap to the boiler, by means of a suitable linkage.

The greatly increased use of gas- and oil-fired boilers has led to more sophisticated control schemes where the control of the heat input can be achieved quickly and accurately.

Small-bore and micro-bore systems

These have been accepted as a much improved method of providing central heating systems for domestic premises. Both systems differ basically from the thermosyphon system in that the water is circulated by pump and therefore eliminates the need for large diameter pipes. In many instances the domestic hot water is still heated from the same boiler on a separate thermosyphon system and therefore is unaffected by any thermostatic control effected on the radiator system. In other cases the domestic hot water heating is part of the pumped system but incorporates separate thermostatic control to ensure that the temperature of the domestic hot water is held at an acceptable level. The boiler is controlled by an electric on/off thermostat, which in turn controls the fuel input by means of a solenoid valve. Generally the thermostat is set to maintain a constant boiler water temperature of approximately 77°C and will maintain this temperature regardless of the various demands made upon the boiler.

On/off control

For a simple background heating installation, thermostatic control of the boiler and/or pump is often sufficient. In the case of full central heating, thermostatic control can be effected in several ways, the simplest being a thermostat sited in a part of the house representing typical temperature.

Another factor that needs special consideration arises when an automatic control system is installed on a solid fuel boiler. These boilers are themselves thermostatically controlled by means of a thermally-operated flap or fan working in relation to the temperature of the water passing through the boiler. If a thermostatically-controlled central-heating circuit is demanding full heat under adverse conditions, the boiler thermostat will be wide open in its efforts to produce sufficient air for combustion to maintain water temperature at 82°C. When the room thermostat is satisfied, it will switch off the circulator and cause all flow of water through the boiler to cease; a rapid rise in water temperature in the boiler will then ensue. The thermostat on the boiler will close the air supply to the fire as the water temperature rises and it is for this reason that it is recommended that thermostats on boilers being used for this type of heating circuit have a maximum differential or throttling curve of 5°–6°C.

The residual heat in the incandescent fuel within the boiler will still be considerable, and this is dissipated through the separate thermosyphon system supplying the domestic hot water. In most circuits where the room thermostat only switches off at infrequent intervals this arrangement works quite satisfactorily, the domestic hot-water tank acting as a thermal buffer and accepting the unwanted heat from the boiler at the point where the thermostat switches off. However, on a circuit where the thermostat cycles frequently this could lead to some difficulties especially if the demand for domestic hot water was low, for after two or three cycles in a short space of time the domestic hot-water temperature could build up to a dangerous level.

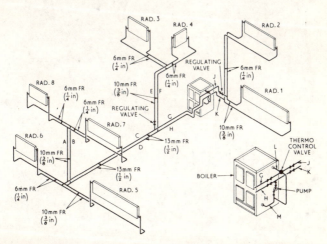

Fig. 10.5(a). Arrangement of two-pipe small-bore system; cold-water feed and domestic cylinder not shown.

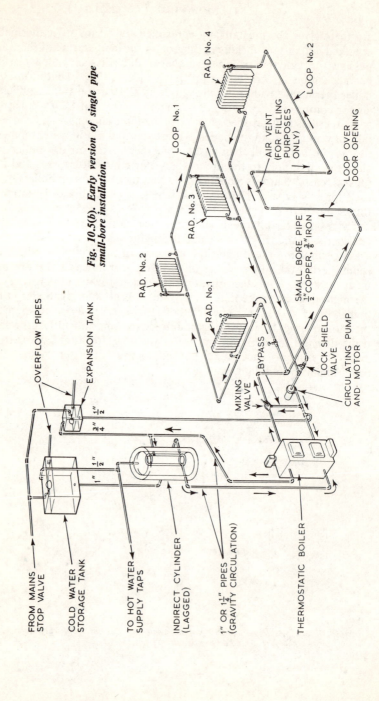

Fig. 10.5(b). Early version of single pipe small-bore installation.

FROM MAINS STOP VALVE

OVERFLOW PIPES

COLD WATER STORAGE TANK

EXPANSION TANK

TO HOT WATER SUPPLY TAPS

INDIRECT CYLINDER (LAGGED)

1" OR 1¼" PIPES (GRAVITY CIRCULATION)

THERMOSTATIC BOILER

MIXING VALVE

BYPASS

LOCK SHIELD VALVE

CIRCULATING PUMP AND MOTOR

SMALL BORE PIPE ½" COPPER, ⅜" IRON

RAD. No. 1

RAD. No. 2

RAD. No. 3

RAD. No. 4

LOOP No. 1

LOOP No. 2

AIR VENT (FOR FILLING PURPOSES ONLY)

LOOP OVER DOOR OPENING

½"

¾"

1"

Where the heating boiler burns gas or oil this condition does not arise for the room thermostat could be so arranged that as it was satisfied it switched off both the circulator and the fuel supply to the boiler. With a gas boiler, it could be arranged that a solenoid valve either in the gas supply or the weep of the relay valve would act as a cut-off; with an oil-fired boiler the thermostat would be switched in circuit with the burner and act as an over-ride on the normal boiler thermostat.

Three-way mixing valve

In other cases more refined control is achieved by providing a variation in water temperature through the radiator circuits in relation to outdoor conditions. This is generally achieved by siting a three-way valve in the flow to the radiator circuit and incorporating a mixing device whereby high-temperature water from the boiler is mixed with low-temperature water from the return pipe to give a predetermined flow-water temperature.

The layout of a typical system is shown in Fig. 10.5(*b*). It can be seen that water is circulated through the system by means of the pump which forces water through the boiler and through the flow pipe round the system. Between the circulating pump and the boiler a bypass is fitted, so that if necessary the circulating pump can force water through this alternative route and bypass the boiler. At the point where the bypass pipe joins the main flow a three-way valve is fitted so that the ratio of water being dispatched round the hot-water circuit can be varied between 100 per cent from the boiler to 100 per cent re-circulated water from the return pipe. This three-way mixing valve can take one of the following forms:

(1) A manually operated device. This is set by trial and error to produce a flow of temperature calculated to maintain the room conditions at a desired temperature.
(2) A thermostatically controlled mixing valve which once set to a temperature will vary the ratio of selection from the boiler or return pipe to maintain a constant flow temperature.
(3) A thermostatic selecting device so arranged that through a sensing phial sited outside the building it can adjust the flow temperature in relation to outside temperature.

The first method of manual control is somewhat problematic in its performance and its final effect on the efficiency of the system. Method (2) uses a self-contained thermostatic three-way mixing valve, generally as shown in Fig. 10.6, where it will be seen to consist of a three-ported valve, the outlet being permanently open and the other two ports being

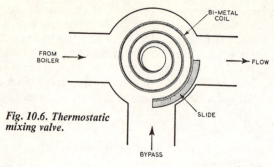

Fig. 10.6. Thermostatic mixing valve.

controllable through the action of the thermal element. This type of control has not been widely used until the present, although several patents have been taken out recently which suggests that more sophisticated versions might become available in the next few years. This method still leaves the householder to adjust the flow temperature manually according to changes in outside temperature, but once set would not require any further adjustments until the outside temperature dictated this.

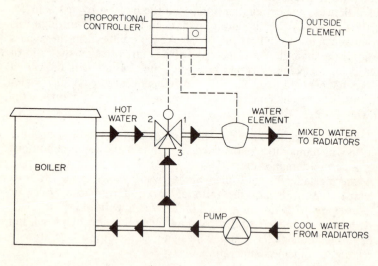

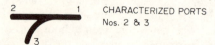

Fig. 10.7. Inside/outside control system. (Drayton.)

Method (3), which achieves automatic adjustment of flow temperature in relation to outside temperature, is becoming increasingly popular. In its early development this type of control was achieved hydrostatically by a self-contained mixing and sensing device, and whilst this was very effective, its complication and limited adjustment made it somewhat restricted in its application. With the advent of solid state controls the same principle of operation has been achieved in a much more sophisticated manner. Fig. 10.7 shows the basic principles on which such a control operates. The mixing valve is generally sited as shown on the flow pipe from the boiler, taking its further connexion from the return pipe downstream of the pump but upstream of the boiler. The mixing valve comprises a three-ported valve with an electric motor controlling the position of the porting and capable of being operated in a series of steps. The outdoor sensing is of the electrical resistance type and changes of temperature send a signal into the control amplifier, which repositions the mixing valve in relation to changes of temperature. A sensor is also sited in the flow pipe to the radiator circuit and changes of temperature in the flow, as effected by the position of the mixing valve, cause signals to be sent back to the amplifier control. By means of an electronic form of Wheatstone bridge circuit, the system is arranged to move the mixing valve to a position whereby the signals from the outdoor- and flow-sensing devices are balanced out and the correct ratio of outside temperature and radiator flow temperature is achieved. Figure 10.8 shows a typical graph of the desired relationship between flow temperature and outside temperature suitable for ideal application conditions.

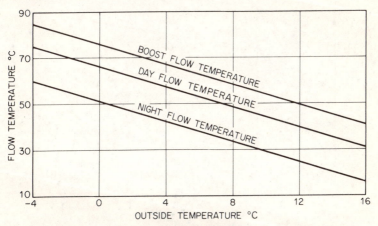

Fig. 10.8. Flow control effected in relation to changing outside temperatures.

As each installation varies from one to another in relation to the position of the house, solar gain, boiler size and its effect on rate of change of temperature in the building, it is desirable and sometimes necessary to depart from the conditions shown in the graph in

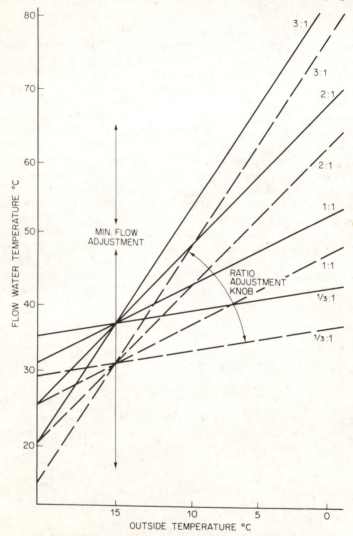

Fig. 10.9. *Effect of adjustments on (a) minimum flow temperature and (b) temperature ratio. Drayton Theta inside/outside control. (Drayton.)*

Fig. 10.8. With electronic control such adjustments can be provided, and Fig. 10.9 shows a typical graph of the type of adjustment that can be achieved. The two important adjustments are outside air/water temperature ratio and the ability to lift this ratio bodily up or down, both of which are achievable by two separate adjustments.

In addition to this method of controlling small-bore and micro-bore circuits, other systems can be used, including (1) separate radiator valves for each radiator and (2) zone valves controlling individual radiator circuits.

Radiator valve control

The use of individual thermostatic radiator valves on central heating offers some flexibility, as in this way individual requirements for each room can be made comparatively easily. The general principle employed in all thermostatic radiator valves comprises a characterized valve head which can progressively throttle the supply of water to the radiator. This valve head is moved by a thermal sensing device, located either in the head of the valve or remotely sited in the room. Generally, a means of selecting operating temperatures is provided, so that once set, the radiator temperature will rise or fall in relation to changes in room temperature. Of the many valves available, Fig. 10.10 shows a typical example. In this model the initial setting is achieved by rotating a knob to a suitable position on the range scale. This action moves the sensing element and valve further away from or nearer to the valve seat, thus determining the amount of thermal movement that must take place to close the valve. Under normal working conditions the valve will be wide open at low temperature and closes as the temperature rises, thus restricting the flow of hot water until a point is reached where the heat supplied from the flow of water into the radiator is just sufficient to maintain the required air temperature. Any change in room temperature will be transmitted to the liquid in the sensing bellows and corresponding change in the rate of water flow will take place.

As in all devices using hydraulic expansion, provision has to be made for over-run conditions, i.e. when the range knob is adjusted for a temperature lower than the temperature prevailing at the moment of adjustment. To meet this requirement the regulating bellows is enclosed by a cylinder connected with the handle, the cylinder being axially displaceable and loaded with a powerful spring. If the valve is completely closed and then set to a lower scale temperature, the cylinder and handle will move outwards. When the temperature of the room goes down to the value as set on the scale, the cylinder will again be forced into its bottom position by the spring and further decreases in temperature will cause the valve to lift from its seat and from then onwards resume normal operation at the lower temperature.

As with all proportional devices, some degree of offset can be encountered if the heat requirements of a room in which the radiator is situated change drastically. This would especially apply where the valve

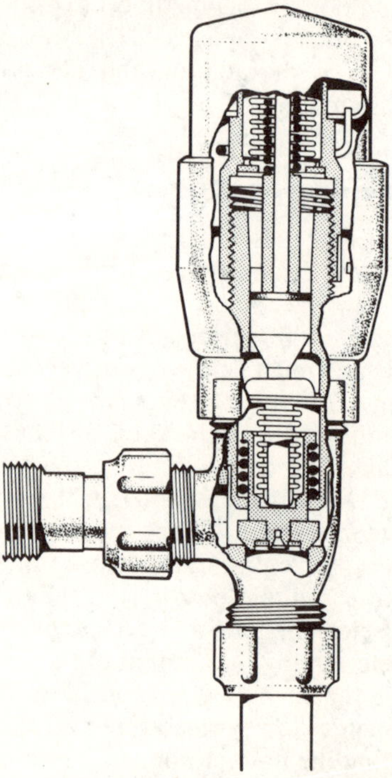

Fig. 10.10. Radiator valve with built-in sensing head. (Drayton.)

was controlling the only radiator and the only source of heat in the room. In actual fact this is rarely the case and it would normally be working in conjunction with other radiators not thermostatically controlled. In these instances the thermostatically controlled radiators are left permanently on, providing the basic heat input, the manually controlled radiators being brought in when the heat demands are at a maximum. In this way, close room control can be obtained and the offset that is associated with full proportional control can largely be avoided.

Thermostatic valves are available, which offer various means of achieving temperature control, two of which are shown in Fig. 10.11. In these instances, instead of having a temperature selector incorporated into the head of the thermostatic valve, it is sited remotely and thus is able to meet situations where it is neither convenient nor technically accurate to have a radiator controlled from a temperature sensing device close to it.

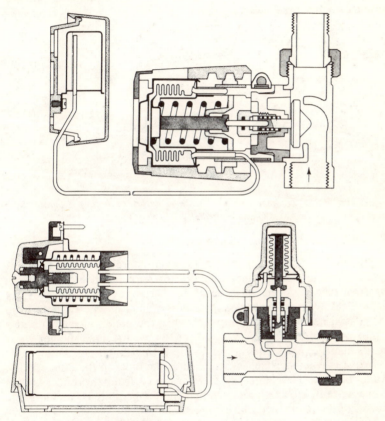

Fig. 10.11. Radiator valves with (top) remote sensor and (bottom) remote sensor and temperature adjustment. (Dunfoss Ltd.)

Zone control

On larger domestic heating schemes, it is sometimes desirable to provide separate temperature control for several parts of the same basic system. Such a condition could arise where all the bedrooms are to be maintained at a lower temperature than the ground floor or where one side of the building is subject to much higher heat losses or solar gain than the other. In such instances the radiator circuits are divided and each circuit is provided with a separate valve, normally operated by a room thermostat. Such a circuit is shown in Fig. 10.12.

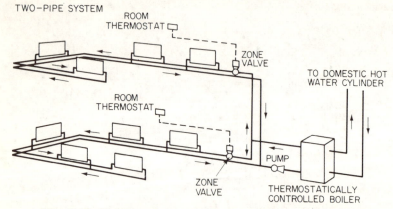

Fig. 10.12. *Two-pipe system with zone valves controlling separate circuits. (Satchwell Controls Ltd.)*

Selective control for domestic hot water

In many schemes, the domestic hot water is provided from the same boiler that supplies the hot-water radiator circuits. The domestic hot water is then supplied via a separate thermosyphon system so that domestic hot water is provided regardless of intermittent demands from the central heating circuit. Even so, when the boiler is being run at a temperature of 77°C, it is inconvenient to have the domestic hot water at a similarly high temperature. To meet this condition, thermostatic valves are available which, when sited in the pipes to the domestic hot water tank, will regulate the flow of water from the boiler to the domestic tank in such a manner that the temperature of the domestic hot water is maintained at a comfortable level.

It is increasing practice to supply the domestic hot water from the pumped circuit, and in cases where it is desirable to have separate control of the domestic hot water a two-position diverting valve is used. This is sited on the flow pipe from the boiler and is so arranged that it directs the flow either through the radiator circuits or through the domestic hot water circuit. Two thermostats are used; the first is a clamped-on thermostat fitted to the domestic hot water tank and the control is so arranged that whenever the domestic hot water is below the required temperature the diverter valve supplies hot water from the boiler to the domestic system until this is satisfied and the clamped-on thermostat switches the diverter valve to send the boiler output of hot water through the radiator circuit. Under these conditions the second thermostat, which is a room temperature detector, controls the boiler on/off in relation to the air temperature in the room. When this is satisfied then the boiler and the circulating pump are switched off. An alternative method of connecting the controls can be used to give priority to the heating circuit instead of the hot water circuit.

Low-capacity boilers

Most of the schemes already described are based on the use of a conventional boiler having a cast iron heat exchanger of comparatively high thermal mass. In such designs, on/off control is satisfactory in that any slight over-run of temperature caused by this method of control is safely consumed within the thermal mass. Another advantage of on/off control is that the boiler is always operated at full gas or oil flow during the on cycle at maximum efficiency, avoiding the lower efficiency encountered if the boiler is run at anything other than full fuel rate. In recent years considerable advances have been made in the use of heat exchangers having very low capacity and low thermal mass. These can consist of stainless-steel heat exchangers where the waterways consist of pipes and thin-walled channels. With this type of construction high efficiencies can be achieved at heat inputs other than maximum. It is essential with this type of construction to avoid any tendency for temperatures to over-run the desired condition. To meet both of these conditions it is convenient to arrange for the fuel input (generally gas) to be modulated rather than controlled in an on/off manner. This has given rise to a number of thermostatic controls where the water temperature can be modulated to suit outside conditions and therefore maintain the boiler in a continuously running state, except when all conditions are satisfied and the fuel rate is turned to a pilot rate or is turned off completely.

HEATING BY MEANS OF HOT AIR

In addition to the schemes already described there is an increasing interest in heating domestic and commercial buildings by ducted air from a centrally-sited heat exchanger. Two basic types are in current use, one obtaining its heat from a directly-fired furnace or heat exchanger, and the other from a water heater which supplies hot water to a heat exchanger through which air is forced by fan and is then ducted to various parts of the building.

Directly-fired heat exchanger

In the first system, where the heat exchanger is warmed directly by gas or oil burners, the heat exchanger can consist of a series of tubes through which the products of combustion are fed and across which is

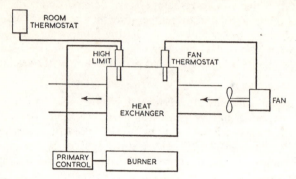

Fig. 10.13. Diagrammatic arrangement of controls for a warm air burner.

blown the air to be heated for warming purposes. The basic control requirements are shown in Fig. 10.13, from which it will be seen that the burner depends for its efficient functioning on the normal primary controls. These will include a flame-failure device, an excess temperature device in the flue, and the usual limit thermostats applicable to the primary control.

The control applied to the output of the heat exchanger begins with the room thermostat. On a fall in temperature this calls for heat and, if the high-limit thermostat is in the closed position, the signal will be passed to the primary control which will set the burner in motion. When ignition is established, the heat exchanger will begin to warm up and the fan thermostat will switch in the fan.

The high-limit and fan thermostats are complementary to each other and several types of control are available in which both their functions are built into one system operating from a common thermal element. With this arrangement it is possible to ensure that there is a minimum difference between the setting temperatures of the high-limit and fan switches. This is necessary because after the fan has been started it must be allowed to run for a certain time to enable it to pass air through the heat exchanger and so stabilize the temperature conditions in the exchanger. If the high-limit thermostat was set to operate too closely to the point where the fan switch cut in, it would be possible for a slight temperature rise in the heat exchanger to continue immediately after the fan switched on and cause the high-limit thermostat to cut out unnecessarily.

The high-limit switch is usually arranged to have a lock out device so that once it operates it requires manual resetting to set the normal control system in action again. The settings at which these two switches operate vary from one application to another, and will depend largely on the thermal mass of the heat exchanger. An exchanger having a low thermal mass will react quickly to the heat input from the burner and its temperature will rise rapidly as soon as heat is available. In such cases the setting temperature of the high-limit thermostat would need to be appreciably above that of the fan switch to prevent unnecessary locking out. On the other hand a heat exchanger having high thermal mass will react slowly to the heat input, and as soon as the fan switches on, the rise in temperature in the heat exchanger should be arrested. In such cases the high-limit thermostat need only be set at a slightly higher temperature than the fan switch.

The above control system describes the minimum requirements for this type of heat exchanger. Many variations are available including two-step control of the burner giving two rates of heat input, and two-stage control of the fan producing two fan speeds. Both these additions provide a refined control on the final output of the heat exchanger and on the control of temperature in the rooms being heated by the hot air.

In small domestic versions of this type of heating, thermostatic control is normally limited to on-off control by the thermostat; separate heating requirements for the rooms being supplied with hot air are catered for by sharing the output from the heat exchanger through the medium of adjustable outlet grilles at the point of discharge.

An alternative or supplementary arrangement to this scheme utilizes motorized damper units built into the grille outlet. This takes the form of a small motor connected to a damper flap behind the grille, and is switched either to the open or closed position by a room thermostat. In this way the hot air output from the heat exchanger can be controlled into each room, thus compensating for various heat demands in different sections of the building. Obviously, each motor is preferably controlled by a thermostat sited in the room into which the flap-controlled grille outlet discharges.

Indirect heat exchanger

The second type of heat exchanger, whereby the heat is supplied in the first place through a hot-water boiler, needs only a simplified version of the above controlled scheme. The water supplied to the heat exchanger is controlled in temperature through the boiler thermostat. The fan supplying the air across the heat exchanger can be controlled by a room thermostat sited in a principal room, or in a position where average house-temperature conditions can be sensed.

11

COMMERCIAL AND INDUSTRIAL HEATING CONTROLS

THE automatic control of large central-heating circuits follows the same basic principles as for domestic circuits, the general difference being that the larger schemes offer more scope for fully automatic control which, in turn, gives scope for a wide variety of control methods.

HEATING BY HOT WATER AND LOW-PRESSURE STEAM

In all large schemes the source of heat, whether this be a gas, oil, or solid-fuel boiler, calls for exact control of water temperature. Thus, the appliance itself must have its own control equipment to ensure an adequate supply of hot water at any temperature demanded by the hot-water circuit. The majority of schemes arrange that the output from the boiler should be maintained constantly at the top temperature requirement, say 80°C. From then onwards the distribution of the water can be controlled by several different methods of which one uses inside-outside temperature-compensated mixing valves (Fig. 11.1).

Zone heat control

In large heating schemes where the one hot-water circuit has to meet the needs of all parts of the building (which may suffer from various influences such as solar heat gain in relation to the position of the sun) it is sometimes necessary to split the heating circuit into several zones each of them having their own indoor–outdoor control. In this case the boiler supplies water to two or three mixing-valve circuits, each one receiving a signal from its inside-outside compensator system and providing the correct flow temperature for the particular zone which it serves (Fig. 11.2).

The provision of water at a temperature in relation to outside conditions will set the basic heat input into the zone being heated. In some cases this control is sufficient to maintain acceptable temperatures. However, it does not take into account load factors within the building which may disturb the averaging produced by the compensating system; and some over-riding control in the form of room thermostats

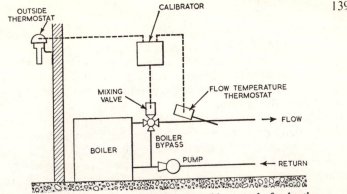

Fig. 11.1. Essential components for inside-outside control of a heating boiler.

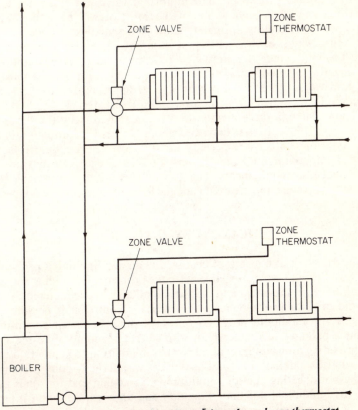

Fig. 11.2. Individual zone control using mixing valve and zone thermostat. Each zone would normally have its own pump installed in its return pipe.

is therefore sometimes needed. This can occur for instance when buildings are used for public meetings and the temperature in a room can rise rapidly when it is full of people. Where room thermostats are fitted, they are usually an integral part of the control system and have a proportional effect on the flow temperature of the water rather than over-riding on-off control.

In some circuits the inside-outside compensating control is dispensed with completely, and all heating is controlled through zoning thermostats. This would apply where outside conditions have less influence on the internal conditions, which might be such as to create widely different heat requirements. In these instances the water aviable from the heating source is maintained at a constant temperature, and a motorized valve attached to each zone is modulated in relation to the demands from the zoning thermostat. The motorized valve will either act as a throttling valve on the flow line of the zone, or will be a three-way mixing valve circulating through the flow pipe a mixture of boiler water and return water. In these circuits it is important to place the thermostat at a point where it is most able to sample the average conditions of the zone. In some instances this is not possible with one thermostat and it is necessary to install two or three arranged to average between them the final control effected on the motorized valve.

Finally, there is the individual control of radiators by the thermostatic valve similar to that described under Domestic Heating Systems. This type of control is especially useful where the building is subdivided into numerous small rooms, or offices, each one having its individual heat requirements. This type of control also overcomes the difficulty of reconciling different personal ideas on comfortable temperatures.

HEATING BY HOT AIR

The use of air as a means of heating and air conditioning is probably the most flexible system available when the problem of providing correct conditions in commercial and industrial buildings is encountered. With the aid of suitable equipment and automatic control it is possible to provide any conditions ranging from: (1) tempering the air conditions within a building, mixing fresh air with a certain amount of heated air; (2) providing all heating required so that in addition to ventilation the heated air counteracts the total heat losses within the building; (3) providing both heating and air conditioning, arranging that the air is put into the building at a definite humidity and temperature; and (4) providing refrigerated air in summer conditions where heat gains within the building can be counteracted by air provided at a low temperature.

Such a subject involves such a wide variety of control problems that it is well beyond the scope of this book. It is sufficient to give an outline of the basic control circuit associated with hot-air heating systems without delving into the more complicated control problems associated with full air conditioning.

Forced warm-air ventilation

Heating through forced-air circulation systems can provide either heated, ventilating air which will provide the air requirements for a building, or with additional heat can contribute all or part of the actual heat requirements, replacing the heat losses in the building. Considering the first problem, that of merely taking the chill off fresh air, the method of control is comparatively simple and is illustrated in Fig. 11.3. Here it will be seen that air is drawn from the outside, is passed

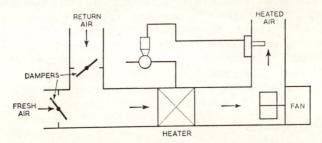

Fig. 11.3. Diagrammatic arrangement of controls for air tempering on ventilation equipment.

through a heater and, via a fan, is discharged into the building. A second duct takes a percentage of return air and, by manually adjusting the dampers on the return air and outdoor air ducting, a fixed percentage of each source can be drawn into the building.

The heater may take the form of hot water or steam coils positioned in the main trunking so that the air is drawn across it by the fan. The supply to the heater is controlled through a modulating valve which, in turn, is controlled by a thermostat in the discharge air. The thermostat and valve form a modulating control of either the electric or electronic proportioning arrangements described in an earlier chapter.

In selecting the thermostat and the valve for this type of control probably the most important factor is to ensure that the rating of the heater and the valve size are exactly right for the possible demands likely to be made on it. In the first place the heater must be sized so that it is only just large enough to cope with the worst conditions likely to apply, and secondly, the selection of modulating valve must be such that its throughput, whether this is steam or hot water, is only sufficient to meet the demands of the heater under full load conditions. If the valve is undersized then even when the thermostat is demanding full heat there will be insufficient temperature rise as the air passes across the heater. On the other hand if the valve is oversized the modulating action of the thermostat will be partially lost and something approaching on-off control will be achieved. This will be due to the fact that on a rise in temperature the thermostat will cause the valve to modulate towards the closed position, but, being oversized, the valve will use most of its travel before any effective control begins to take place on the heat input to the heater. Thus, instead of taking full advantage of the modulating action of the valve, it will only begin to affect the heat output from the heater as it reaches its near closed position.

Another factor which will contribute to the success, or failure, of such a system is the sensitivity of the thermostat itself. With a high rate of air discharge, any change in the heat input caused by the heater being throttled or opened by the modulating valve will have rapid action on the air temperature, and unless the thermostat is sufficiently sensitive to keep pace with this change a continuous hunting cycle will ensue, with the production of rapidly rising and falling temperatures on the air discharge in its attempt to maintain a mean temperature.

Hot air heating

Where this type of system is expected to provide the basic heating for the building the main requirements are that the heater should be sufficiently sized to provide the heat required, and that suitable automatic controls are provided so that a constant temperature is maintained within the building. The control system, therefore, is as shown in Fig. 11.4, with a thermostat working on room temperature. The first control in the discharge duct now merely acts as a limit thermostat to ensure that the temperature of the air being discharged into the room does not rise above, or fall below, a temperature which would cause discomfort at the discharge points. The controlling thermostat sampling room conditions can either be in the building or in the return air duct, and will act as the primary controller of the motorized valve on the heater in the ducting. Thus, as room temperatures rise the thermostat in the return air or in the building will sense this and transmit a signal to the motorized valve causing this to move towards the closed position, and so reduce the heat input into the building. The thermostat in the discharge duct will monitor this reduced temperature to a point where, should the outgoing air drop below a certain comfort

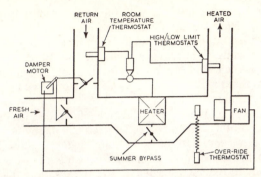

Fig. 11.4. Basic control circuit for heating by hot air.

temperature, it will over-ride the room or return air thermostat and limit the travel of the modulating valve on the heater so that the discharge air is kept above a predetermined minimum temperature. Similarly, the high-limit thermostat can be housed in the discharge air duct so that where the room, or return air thermostat, is calling for additional heat the motorized valve will travel towards the open position but, should this cause an uncomfortably high discharge air temperature, the high-limit thermostat will limit the travel of the modulating valve to a point where the maximum discharge air temperature is kept within a certain comfort limit.

Protection against freezing

On both these systems, whether it is for ventilation or for heating, the heating coils themselves must be protected from freezing in the event of the heat being switched off and the fan left on, or if for some other reason cold air from the outside is drawn across the heater causing it to freeze. The simplest form of protection is to have a motorized damper operating the incoming air duct and controlled by a thermostat between the fresh air duct and the heater. This is set so that if the air entering the heater drops to freezing point the fresh air damper goes to the closed position, shutting off the air from the outside. The same damper motor may be arranged to operate the return air dampers so that as it closes the fresh air inlet it proportionally opens the return air inlet.

Control of electric heaters

The systems described above utilize a motorized valve modulating the supply of hot water or steam to the heater coil. An equally satisfactory control can be obtained with the use of electric heaters where the heating elements are sub-divided into a series of steps, and these steps are switched on, or off, through a motorized shaft driving a cam, or a series of cams, connecting switches to the steps in the electric heater. Thus as the thermostat modulates the motor position, the cams are rotated so that they progressively switch out, or switch in, the steps in the electric heaters.

As alternatives to multi-step switching of electric heaters, control circuits incorporating silicon-controlled rectifiers or thyristors can now be used. The advent of these solid state switches allows the whole heating load to be switched on and off continuously at a sufficiently high frequency to reduce to negligible size the resulting fluctuations of the leaving-air temperature. At any instant the load is either zero or 100 per cent, but the temperature rise depends on the mean load, i.e. the percentage of the time the load is switched on. The percentage 'on' time can be continuously varied, so that fully modulating control of temperature can be obtained. Multi-step control cannot give comparable results unless the number of steps is made very large.

UNIT HEATERS

Unit heaters are used in increasing quantities for heating shops and commercial and industrial premises. With its ability to supply quick heat directionally, the unit heater offers a means of providing comfort conditions in large spaces where it is impracticable or unnecessary to heat the whole building. Most of these heaters comprise either a series of hot-water or low-pressure-steam pipes, or electric elements, across which air is driven by a fan and heated. The air leaves the unit heater at a fairly high velocity and can be directed towards the zone to be heated.

The method of applying thermostatic control to unit heaters will depend on the number of heaters employed and the degree of accurate temperature control required. The simplest method is the two-step on-off control where a room-temperature thermostat is linked directly to the fan, which is switched on and off according to the demand for heat. In the steam-heated unit heater the steam is left on continuously, whilst in the electric heater the same room thermostat also controls the elements through a contactor; a high-temperature excess switch being provided so that in the event of fan failure the heater elements are switched off as their temperature rises owing to the lack of air passing across them. Suitable interlocking switches are normally provided on electric heaters so that for summer use the heating elements may be locked-out although the fans continue to operate to provide air circulation.

In the steam unit heater controlled through on-off switching, there is generally a low-limit thermostat so arranged that should the room thermostat call for heat when no steam is available at the heaters the low-limit thermostat over-rides the call and prevents the fan from being switched on and forcing cold air into the room being heated.

Modulating control

As an alternative to on-off or two-step control, modulating control offers a more refined and predictable method of controlling unit heaters. Electric, water, and steam-heated units can be controlled in this manner.

The hot water or steam-heated unit presents the easier problems, and in its simplest form can be made to provide modulating control through a direct-acting thermostatic valve (Fig. 11.5). This valve has already

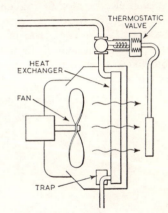

Fig. 11.5. Unit heater controlled by direct acting thermostatic valve.

been described under the general heading of Proportional Controls (see Chapters 6 and 15). The fan is run continuously and should this be required for ventilation only in summer then the steam to the unit heater is shut off manually. Thermostatic control of unit heaters by means of the self-contained valve is economical and convenient, but is somewhat restricted in its application because of the limited length of capillary tubing that can be supplied between the sensing element and the valve, and because the throttling curve of the valve is somewhat coarse in relation to the heat requirements for human comfort conditions.

To overcome these limitations electric modulation is sometimes used. The motorized valve on the steam supply receives its signal from a thermostat remotely sited in the zone being heated, and, through a potentiometric circuit, a change of heat round the thermostat causes the thermal element to move a wiper on the potentiometer creating a state of unbalance in a Wheatstone bridge. This system has already been illustrated in Chapter 8.

Low-limit thermostats are generally incorporated so that in the event of steam or water failure the fan is locked-out, preventing cold air being blown into the zone.

Electric heaters can be provided with some degree of modulating or multi-step control. In these instances the heating element is subdivided into several steps and a multi-step thermostat is supplied so that temperature changes round it cause it to make or break a series of steps each one switching a section of the heater element. Similarly the potentiometric circuit can be employed where a potentiometric thermostat operates a motor to which a series of cams are attached, each cam switching in a section of the electric heaters at predetermined points. On both types of electric modulating control an over-riding high-temperature switch is provided so that in the event of fan failure the high-limit thermostat senses the rapid rise in temperature of the elements and locks-out all the heater until the fan is restored. Modern installations are more likely to incorporate the use of on–off control using solid state switching as described on p. 144.

12

ELECTRIC STORAGE
HEATING

THE use of electric storage heating in commercial and domestic premises
has greatly increased during the last 10 years. Its primary advantage is
that in storing heat obtained during the off-peak tariff periods and
discharging this heat during the daylight hours, it offers considerable
economies over any other form of direct heating by electricity. The most
popular form of storage heating, i.e. the independent unit storage
radiator, has been described in Chapter 14, and this version, together
with the fan-assisted version of block storage heaters, dominates the
domestic market in electric heating.

Other methods of storing electric heat include the Electricaire system
(which is an extension of the storage radiator), hot water storage,
under-floor heating, and the prospect of the increased use of thermal
storage walls.

Whilst the demand for thermostatic control for these various types of
storage heating varies, the basic requirements are quite similar. These
include:

(1) A means of controlling the maximum temperature of the storage
    media.
(2) Either control of the heat input during the charge period,
    preferably in relation to the expected temperatures of the outside
    environment during the discharge period, or, control of the output
    from the storage media during the discharge period, as an alterna-
    tive to or supplementing the controlled charging rate in the
    former requirement.

Application differences vary these basic requirements to some degree.

Electricaire

This is basically an extension to the fan-assisted storage radiator,
described in Chapter 14, with the exception that whereas in the case of
a single radiator air is discharged into the room in which the radiator
is sited, the Electricaire system allows for the ducting of hot air to the
various parts of the building. A thermostat is provided for the control

of the input of heat into the storage media limiting this to a certain maximum temperature. This usually takes the form of an expanding tube/core thermostat, which is actually inserted into the heater bricks and so gives a more accurate control of core temperatures than the less expensive bi-metal versions used in unit storage radiators. Where this system of heating is employed on low-cost night tariffs, provision is made to top up the stored energy during the daytime if the output air temperature in the ducting falls below a given value. This is arranged by siting a thermostat in the outlet air ducting, which will switch on a percentage of the elements in the storage medium on a fall in temperature. Alternatively it can be arranged that when the air discharge temperature during the daytime falls below the given value, this can switch on separate direct-acting elements, through which the discharged air passes, and so raise the temperature to an acceptable level. In any of these systems where power can be called for at any time other than the off-peak period, the circuitry is so arranged that these additional calls for power are registered on the normal charge meter and not on the low-tariff meter.

Hot-water thermal storage

In this system heat is stored in a large amount of water during the off-peak period and then arranged to slowly dissipate its heat through a conventional hot-water heating system with radiators, etc. The problems of storing large amounts of high-temperature hot water and suitably insulating it against unwanted heat losses have up to now precluded its use in large quantities in normal dwellings. Water is heated up to its safe temperature below boiling point; sometimes the system is subjected to an artificial pressure head, thus elevating the boiling point up to 150°C. The control thermostats employed are generally set to operate at some 5°C below boiling point. As a protection against control thermostat failure, a pressure switch and safety valve are provided, as with other hot water systems. Once having heated the stored water the control of its use in relation to the temperature of the rooms being heated is almost identical to that described in Chapter 14, covering the inside-outside control associated with conventional central heating systems. Here water is circulated through the radiators and a mixing valve takes a percentage of high temperature water, mixing this with the return water and recirculating it through the radiators at a temperature dictated by outside air conditions.

Under-floor heating

The use of under-floor heating is now well-established practice, both in domestic and commercial premises, and consists of heating cables embedded in the concrete flooring using this as the heat storage media. Experience has shown that a loading of 150 W/m² gives an acceptable surface temperature between 24° and 27°C, this being the maximum temperature for comfort. Under these conditions uniform air temperatures throughout a room can be maintained, and experience shows that little stratification occurs between floor and ceiling. The temperature control of this system at first presented some problems, in that the thermal inertia between the charge going into the floor and the point where any room thermostat registers as being satisfied could mean that long after the thermostat has switched off, the heat stored would continue to discharge and cause further rises in temperature. The other problem was of gauging the heat to be stored during the off-peak period to give an acceptable room temperature during the following day. Even so, early systems operated reasonably successfully with a simple control system consisting of a room thermostat and a separate limit thermostat embedded in the floor and arranged to switch off the heating if the floor surface temperature rose above the comfortable maximum of 27°C.

Floor heating, and other forms of heat storage lend themselves to some form of external temperature control system.

External temperature controllers

Many of the larger storage heating systems already mentioned, designed to take advantage of low-cost night tariff, pose a basic control problem in deciding how much thermal charge is required during the night period to provide the correct room temperatures during the following day. Many schemes have been devised to meet this problem, most of them based on a system of measuring the outside temperature during the night and from this, calculating and controlling the stored heat energy required during the night period to provide the correct conditions during the day. A typical system in current use, the Proscontroller, can be described as follows:

This system (Fig. 12.1) automatically proportions the charge to the heating system throughout the whole of the available low tariff period according to the difference between the external temperature and the desired internal temperature. It consists of two items, the control unit, containing a thermostat and heating coil (Fig. 12.1(a)) which is suitably weatherproofed for use on the outside of the building, and the calibrator unit (Fig. 12.1(b)) sited inside the building and containing an auto-transformer and variable resistance.

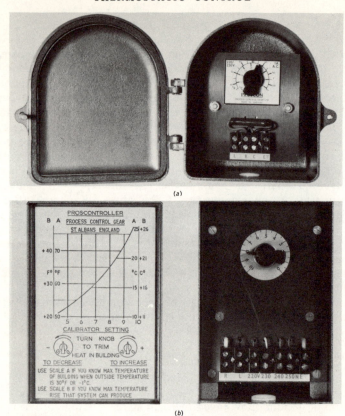

Fig. 12.1. *External temperature controller. (a) Thermostat and heating coil and (b) calibrator unit. (Process Control Gear Ltd.)*

The loading of the heating coil in the control unit is adjusted so that the same temperature rise can be produced in the control unit as the heating system can produce in the building, thus the control unit is virtually a model of the building to be heated but without the building time lag. The setting of the thermostat in the control unit determines the temperature to be maintained in the building, whilst the loading of the heating coil in the calibrator unit determines the proportional band covered. For a typical installation calling for an internal temperature of 18°C to be maintained with external temperatures down to −1°C the percentage charging time will vary over the temperature band, ranging from 100 per cent charge when the outside temperature is −1°C, to nil when the outside temperature reaches 18°C (Fig. 12.2). To meet

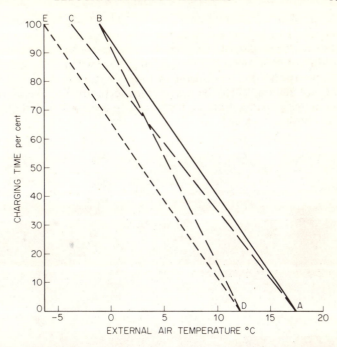

Fig. 12.2. Proscontroller Operating Characteristics. Line AB: typical conditions as described in the text, Line AC: effect of altering calibrator resistance only, Line DE: effect of altering thermostat only and Line DB: effect of altering thermostat and calibrator. (Process Control Gear Ltd.)

such a requirement the resistance of the calibrator is adjusted to a value such that the heating coil in close proximity to the thermostat in the outside control unit just fails to cause the thermostat (which is set to 18°C) to open circuit when the outside temperature is −1°C, thus providing for 100 per cent charge under such temperature conditions. Rises in outside temperature will allow the heating coil to convey more heat to the thermostat, causing it to open circuit and cut off the supply to the heat storage elements and the control-heater coil. As the coil cools under the influence of the outside temperature, the thermostat will again call for heat and reconnect the storage supply. The higher the outside temperature the more time the thermostat will spend in the off position. To meet varying site conditions, adjustments are provided, both for the desired controlled temperature and for the proportional band.

To meet conditions where the normal night storage period would be inadequate for maintaining the required internal temperature under the most adverse conditions, i.e. when outside temperatures drop below −1°C, say, a second control system, virtually a duplicate of the first, can be employed. The two systems are connected in parallel, one controlled by a time switch to give night operation only, and the other essentially effecting day operation only. Daytime energy is pulsed into the installation when the external temperature falls below a preset value, the maximum night energy having been put into the system at just above this value.

13

HEATING APPLIANCE CONTROLS

SOLID FUEL APPLIANCES

MOST solid fuel boilers can be used either to heat water for domestic use or to heat water or raise steam to supply the heating requirements for a building or some type of industrial process. In all these applications it is necessary to provide controls to ensure that operation is in the first place safe and, secondly, that it is economical in use and provides the correct heating service required. It was the disastrous consequences of the ignorance on these matters in the early use of steam boilers that led to the development of many of the first devices for boiler protection, especially in respect of water level and steam pressure. The thermostatic controls in which we are interested in the first place are primarily concerned with the safe and economic running of the appliance itself, and therefore this chapter is confined to their functions and duties. In this respect they are generally referred to as *primary* and *high-limit* controls.

The thermostatic control of the heating process to which the boiler is applied presents other problems which are discussed in Chapters 10, 11, 14 and 16; but, briefly, these (secondary) controls can be proportional or on-off in their operation, and in the case of a heating boiler can work on inside-outside temperature control circuits. Their demand for heat will depend on conditions existing either in the building or the process and may be continuous. It is therefore necessary to ensure that the appliance itself is adequately controlled and protected against abnormal demands beyond its capacity. Where these three control functions are present, viz. primary control, high-limit control and secondary control, it is always important to ensure that they are in series with each other and, therefore, allow either the high-limit or the primary control to exercise authority over the secondary control scheme.

Small domestic and commercial systems

The use of thermostatic control on small commercial and domestic solid-fuel boilers has been established for many years, but until recently has been generally confined to boilers above the 63 MJ size. The thermostatic control generally took the form of a bellows-operated lever connected to the air-inlet flap, a typical example being shown in Fig. 13.1. A long bellows, generally filled with paraffin wax or other hydraulic fluid, was inserted in a pocket built into the water space. On a rise in temperature the liquid in the bellows expanded, causing the bellows to thrust upwards. This tilted the lever and caused the air flap to drop and so reduce, or cut off entirely, the air supply to the fire. This comparatively simple system was used extensively and is still to be seen on many commercial boilers.

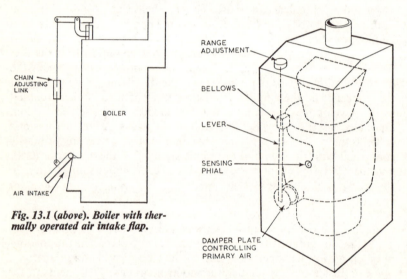

Fig. 13.1 (above). Boiler with thermally operated air intake flap.

Fig. 13.2 (right). Domestic boiler showing damper flap operated by vapour-pressure thermal system.

Thermostatic control of small domestic solid-fuel boilers is a comparatively new innovation, but has already standardized itself generally into the system shown in Fig. 13.2. Here, a vapour-pressure bellows system is employed, the sensing phial being sited in a pocket in the water jacket and the bellows held in a bracket in hard contact with a lever. At the end of the lever is a disc sited in line with the air intake to the boiler. The boiler is constructed so that air is only allowed to enter the combustion area through the one air port, all other flue doors, charging doors, etc., having surfaces ground (and therefore air-tight) fit. As the water temperature rises, the vapour-pressure develops in the bellows and thus moves it forward, tilting the lever and causing the flap to seal off the air intake. A range adjustment is provided in the form of a knob which, when rotated, increases the spring pressure opposing the forward action of the bellows. Thus, when the knob is turned to top temperature maximum spring pressure is developed and the bellows then must develop considerable vapour-pressure before it will move forward.

These devices are calibrated to provide a control over the range 50°C to 80°C. Although the description suggests that the control has an on-off action, in fact the gradual rise or fall of water temperature causes a gradual movement of the bellows and the damper flap. The performance obtained is therefore a modulated action, which enables the damper to be held at an intermediate position so that the air supply to the fire is directly proportioned to maintain the correct water temperature.

One manufacturer has departed from this general pattern, replacing the bellows by a Bourdon tube which moves a damper in the boiler air supply through the medium of a tilting lever. Temperature selection is effected by a range knob which operates on a cam which, in turn, pre-positions the Bourdon tube.

Small gravity-feed boilers with forced draught

In the forced-draught gravity-feed boiler, the fuel is automatically fed from a hopper through a throat of fixed dimensions to the fire bed, the aim being to maintain a constant density and thickness of fuel in the combustion chamber. The combustion air is forced through the fire bed by a fan, the operation of which is controlled by an immersion thermostat in the water jacket of the boiler.

The thermal sensing element can be a bi-metal coil, or a vapour-pressure bulb connected by capillary tubing to a bellows remotely housed in a thermostatic switch. In both types the thermostat controls the boiler in relation to the temperature of the water at its outlet. Such a combination of thermostatic control and forced draught enables the boiler to respond rapidly to heat demand and is very suitable for use on forced circulation systems where thermal lag of the heating appliance (both in reaching temperature on a demand for heat and disposing of residual heat after the demand has ceased) must be as small as possible.

Automatic control of hand-fed boilers

The automatic control of this type of boiler follows the same pattern as that already described for smaller boilers, with the exception that as

the firing rate increases so the size of the appliance and with it the sizes
of the flue and air flaps increase. The power required to move the flaps
in relation to temperature change may therefore be more than a thermal
device can generate for direct operation, and servo-mechanisms can be
employed. Whilst most such systems have largely been replaced with
more sophisticated control, a number of them are still in use and it is
interesting to note how final control was achieved. In the system shown

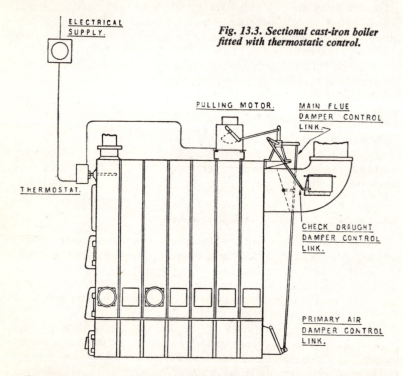

*Fig. 13.3. Sectional cast-iron boiler
fitted with thermostatic control.*

in Fig. 13.3 a single motor is linked by levers to the primary air supply
to the boiler and to the chimney-draught and check-draught dampers.
The linkage is so arranged that as one is opened the other is closed.
Where modulating control is required the thermostat can operate
electrically, or electronically, a potentiometric circuit providing proportional control. Where on-off control is required the thermostat is
generally of the change-over type operating a reversing motor which,
in turn, drives the damper mechanism. Operation of the thermostat
will drive the motor from one extreme position to the other.

The larger automatic solid-fuel boilers have fan-assisted forced draught and automatic boiler-fuel feed in the form of chain grates or worm-drive grates. In both these types of appliance the air for combustion and the fuel are automatically fed into the fire box in relation to the demands for heat as sensed by the thermostat in the water jacket of the boiler.

GAS-HEATING APPLIANCES

Gas-fired water-heating and central-heating units

Gas provides a satisfactory means of heating water, both for domestic hot water and central-heating units. Moreover, it offers a ready means of supplying fully-automatic units capable of working without attention

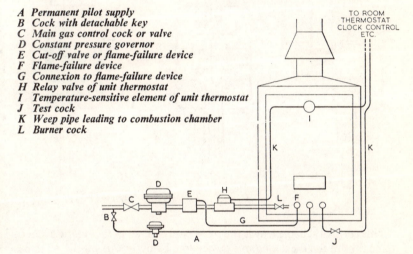

A Permanent pilot supply
B Cock with detachable key
C Main gas control cock or valve
D Constant pressure governor
E Cut-off valve or flame-failure device
F Flame-failure device
G Connexion to flame-failure device
H Relay valve of unit thermostat
I Temperature-sensitive element of unit thermostat
J Test cock
K Weep pipe leading to combustion chamber
L Burner cock

Fig. 13.4. Controls for gas-fired central heating and hot water units.

over long periods. For many years the automatic control of the larger appliances has followed a standard pattern, generally as shown in Fig. 13.4. Here it will be seen that gas flows from the mains via a gas cock *C*, pressure governor *D* and a flame-failure valve *E*, passing through a relay valve *H* which is controlled by the thermostat *I*. This thermostat normally takes the form of a rod-and-tube instrument similar to those described in Chapter 1.

It is arranged that the normal temperature control is effected by presetting the thermostat to the desired water temperature, which may be anything between 50°C and 80°C. On a rise in temperature the expansion of the thermal brass tube causes the valve in the thermostat to close and seal off the small weep supply of gas fed to it from the relay valve *H*. When the weep supply of gas is completely sealed by the thermostat valve closing, the relay valve reacts and closes off the gas supply to the burner. A pilot flame is maintained through the gas cock *B* and through a separate governor *D*. To cover the possibility of flame failure, the safety device is fitted close to the pilot and in the event of flame failure of the pilot the thermo-electric valve *E* cuts off the main gas. This stays off until the pilot is relit, and the thermoelectric valve is manually reset.

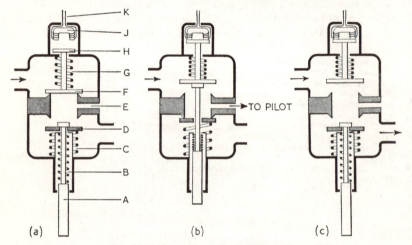

Fig. 13.5. Operation of flame-failure valve.

This control scheme leaves the pilot unprotected in that in the event of flame failure the thermocouple device (see below) shuts off the main gas supply but still leaves the pilot burning. In a preferred scheme, the pilot supply is taken from the flame-failure valve so that in the event of flame failure the pilot too is cut off and remains off with the main gas supply until the valve is manually re-operated and the pilot flame is re-established.

Although the flame-failure device is not a thermostat and is therefore not strictly within the scope of this book, the following brief description will probably be of interest. Fig. 13.5(a) shows the valve with the gas entering from the left and attempting to flow down through the main orifice and out to the burner or pilot from the bottom right; in the position shown, however, the gas is prevented from doing this by the valve *F* being closed and sealing off the main gas supply. Fig. 13.5(b) shows what happens when the button at the base of the instrument is pressed upwards manually, opening the top valve so that gas can flow to a pilot supply from the side but is still prevented from flowing to the main burner. When valve F is pressed upwards by the reset button A, the back of the valve stem is brought in contact with a small electromagnet J, connected to a thermocouple in the path of the pilot flame. As gas is now available to the pilot, this can be lit and the heat from the flame heats up the thermocouple to a point where sufficient e.m.f. is produced by it to energize the magnet and hold valve F in the upper position. At this point the manually-pressed button A can be released, carrying with it valve D. As shown in Fig. 13.5(c) gas is now able to flow through the valve, feeding both the pilot and the main burner. This condition will continue until the pilot light is extinguished, when the thermocouple will cool down and the loss of e.m.f. will de-energize the electromagnet J so that it ceases to grip valve F. The spring G behind the valve F will then close the valve, shutting off all gas, both to the pilot and to the main burner. In some instances, where the gas supply for the appliance is greater than a thermoelectric valve can normally handle, the valve is placed in the weep supply of a relay valve; in this manner it can cut off the flow of gas equal to the capacity of the relay valve and is not restricted by the maximum gas throughput of the thermoelectric device itself.

Modern trends in central-heating requirements call for automatic timing devices, zone controls, and other complications leading to more and more involved control systems for gas water-heating appliances. The trend has been set by the American practice where, instead of having the separate controls as shown on Fig. 13.4, most of them are contained in one control box. The thermostat, being an integral part of the control box, can no longer be a rod-and-tube thermal unit, and is generally either of the liquid-expansion type with a separate sensing phial in the water transmitting movement to a diaphragm or bellows housed within the control box, or an electrically-operated valve.

A typical example of this trend is shown in Fig. 13.6. It consists of a control assembly which can be built up from a series of sub-assemblies. The basic unit consists of the plug-type gas cock at the base of which is a thermocouple flame-failure device for protection of the pilot and for complete cut-off of the main gas supply in the event of pilot-flame failure. The same diecasting that incorporates these features also extends to the left with a top and bottom mounting flange to which can be added various features. In the illustration a gas governor has been added to the bottom flange whilst a low voltage control valve has been added to the top flange capable of receiving a signal from a similar low voltage thermostat mounted at a distance from the control. Other alter-

Fig. 13.6. Multifunction gas control. (Honeywell Controls.)

native features to these include an hydraulically-operated thermostat or a thermopile unit which, through the medium of special thermocouple circuitry, generates sufficient power to operate a small electric valve receiving a signal from a remote thermostat.

Gas-control relay valves

The control of town gas by a direct-acting rod-and-tube or hydraulic thermostatic valve is generally limited to gas flows not exceeding about 1 m³/h. Up to this rate of flow it is reasonably easy to make either of these thermal devices move a valve at a sufficient rate to provide a proportional band capable of controlling from full flow to pilot within some 6°C. Where flows above this figure—for larger boilers—have to be controlled, it is generally arranged that the thermostat controls a small weep supply of gas which, in turn, controls the main gas rate through a relay valve (*H* in the arrangement shown in Fig. 13.4).

This servo-mechanism generally takes the form shown in Fig. 13.7. Essentially this is like a gas-pressure regulator, although its true function is, of course, quite different. Gas flows through the inlet, passes beneath

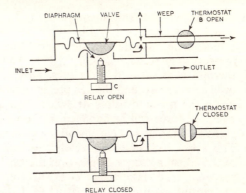

Fig. 13.7. Gas relay valve in open and closed positions.

a valve connected to a diaphragm, and is allowed to escape via the outlet. This condition will continue as long as the pressure of the incoming gas is able to lift the diaphragm and carry the valve away from its seat so that the gas passes on to the burner through the valve, i.e. so long as the pressure of the gas underneath the diaphragm is sufficient to lift both the diaphragm and the weight of the valve. The diaphragm is vented by orifice *A*, and a small amount of gas escapes through this vent up through the weep pipe, through the controlling thermostat *B*, and away either to atmosphere or to the gas burner where the small

Fig. 13.8. Typical solenoid valves for use in conjunction with gas relay valves. Both are suitable for a.c. mains supplies. The built-on case of the right-hand solenoid incorporates a d.c. rectifier for silent operation. (Alexander Controls Ltd.)

gas weep is burnt. Thus so long as the restriction in the weep pipe and the thermostat remains low, the pressure above the diaphragm will be atmospheric, and the pressure under the diaphragm will be the gas pressure. i.e. some 6 m bar higher. This state continues until the thermostat is satisfied and closes. Now no gas can pass through the orifice and away, and the gas pressure above the diaphragm will build up to the pressure below it. When these pressures are equal the unbalanced force will be confined to the weight of the valve, which will then fall on to its seat closing off the gas supply through the relay valve. This action can either completely close off the gas supply or, by means of the adjusting screw C, can shut the gas to the burner down to some predetermined maintenance rate. This condition will continue until the thermostat again calls for heat when it will open and thus release the gas pressure on top of the diaphragm.

The thermostat in the weep supply can take any form, e.g. room thermostat, immersion, or high-limit control, and in addition to this the same weep line can carry a time-operated device (as shown in Fig. 13.4) so that the appliance can be controlled through definite cycles throughout the day.

Where electric controls are used in conjunction with a gas-fired boiler, a solenoid valve inserted in the weep will control the gas supply to the burner. The solenoid valve can be made to operate by a signal from an electric room thermostat, a timing device, or other more complex control schemes. Typical solenoid valves are shown in Fig. 13.8.

OIL BURNER CONTROLS

Fuel oil as received from the refinery only burns with some difficulty and to develop its full heat content it has to be vaporized or 'atomized', and then mixed with a considerable amount of oxygen. In *vaporizing burners* a fine film of oil is heated up in a tray or well to a temperature at which it completely vaporizes. As the vapour rises from the heated surface it is mixed with a predetermined amount of air, and is then ignited. In *atomizing burners* the oil is forced through a very small jet, which splits it into a fine spray of minute droplets. These are allowed to mix with a controlled amount of air fed into the combustion zone by means of a fan; the resultant mixture is then ignited electrically and combustion takes place.

To meet these two quite different methods of oil combustion a range of burners is available; they all need careful control of which thermostatic control plays an important part.

Vaporizing burners

This type of burner is able to burn oil at low rates from 12 MJ/h upwards over a range which would normally be below that of the pressure-jet burner. In this respect it has fulfilled a useful purpose for supplying domestic hot water and central heating for average dwellings. Its principle of operation is to allow oil to be fed by gravity into a burner area where it is heated and vaporized. The rising vapour is allowed to mix with a predetermined amount of air and is then ignited. The two principal types of vaporizing burner are the pot burner and the sleeve or drum burner (Fig. 13.9).

The pot burner, as its name implies, generally takes the form of a steel pot with holes perforated round the side. There is an oil inlet in the base through which oil can be allowed to flow from the control. A float device limits the level of the oil in the pot to approximately 3 mm.

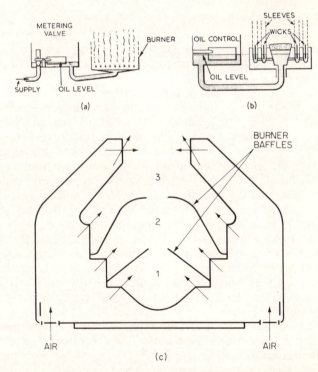

Fig. 13.9. Three types of vaporizing oil burners. (a) Pot burner, (b) drum or sleeve burner and (c) Sesto triple-stage burner showing the three burners within a burner, giving low-, intermediate- and high-fire burning rates.

Once the burner has been ignited the oil vaporizes by the heat obtained from the flame radiation, and thus the vaporizing process is self supporting.

The sleeve or drum-type burner, whilst using the same basic principle, differs in that the oil is contained within one or two annular channels, each having a sleeve rising to some 230 mm above it which, in turn, is perforated to admit air.

The sleeve burner principle is generally confined to the smaller range of burners and has found its place as a conversion burner, mainly for converting solid fuel boilers, cookers and heaters. On the other hand the pot burner has developed in a variety of ways in sizes ranging up to about 23 kW. It can be natural draught, in that all the air necessary for combustion is obtained through the normal chimney or flue pull. On the other hand it can be supplied with a fan, making it a forced draught burner, which becomes virtually independent of any natural chimney pull.

A recent and interesting addition to the range of pot burners is shown in Fig. 13.9(c). It will be seen that three distinct combustion zones are provided in the burner arranging that the oil vapour can burn in a hot environment at all rates of burning and this gives a well-formed and stable blue flame with yellow tips.

Oil flows from the control valve into the base of the burner in chamber 1. The burner can be lit manually by a kindler at the front or the side of the burner or by an electric preheater-igniter element fitted into the burner. Once the burner is heated sufficiently the oil evaporates and, due to the narrow aperture formed by the baffles, vapour spreads along the entire length of the burner and mixes with air from the lower row of air holes. At low fire the oil-air mixture burns partly under the baffle and partly in the aperture of the baffle. The burning gases in the aperture form a blanket, enclosing the lower mixing chamber. As the oil supply is increased, the oil vapour partially mixed with air in chamber 1 escapes to mix further with air entering chamber 2. Again a blanket of burning gases is formed in the aperture of this chamber, now enclosing both chambers 1 and 2. As the oil supply is increased to high fire, the oil-air vapour is lifted into chamber 3 where it mixes with still further combustion air and a characteristic blue flame burns along the length of the burner.

The thermostatic control of both types of burner is fundamentally the same and is an integral part of the flow or primary control. This fulfils three essential functions, namely: (1) to receive oil from the bulk supply and meter it into the burner in such a way that a constant oil level is maintained; (2) to provide a safety mechanism so that if for some unforeseen reason the oil level in the burner begins to rise above normal it will provide a safety trip; (3) to meter the oil into the burner in relation to temperature conditions in the boiler and the demands being made upon it.

Basic oil-control system

The first two functions are not essentially matters of thermostatic control, but as the unit operates as a single entity a brief description of them is included. Fig. 13.10 shows the essential detail of most types of oil control. Oil enters the chamber from the bulk supply, passing through an orifice normally controlled by a needle valve. The position of this valve is controlled by a float in the chamber: as the oil level in the chamber rises the float rises with it and moves the inlet needle on to its seat, so preventing or restricting further flow of oil. The control is mounted at a definite height in relation to the burner, so that the two levels (the level in the burner and the level in the control) are at the desired height. The oil level must be continuously maintained whilst the burner is in operation at the height that will allow the inlet needle to meter oil into the control and out to the burner at the same rate as it is being consumed by the combustion process.

A safety trip mechanism is generally incorporated as shown in the illustrations, so that if foreign matter lodges itself between the inlet needle and its seat and so prevents the valve from closing when normal

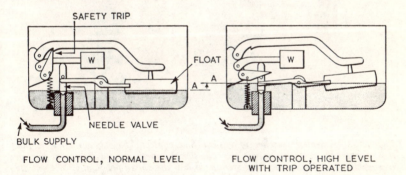

FLOW CONTROL, NORMAL LEVEL   FLOW CONTROL, HIGH LEVEL WITH TRIP OPERATED

Fig. 13.10. Vaporizing burner level control device. Left, normal operating level. Right, rise in level caused by dirt on inlet needle causes float to continue to rise and operate mousetrap type trip arm which enables spring loaded cam to hammer inlet needle into closed position.

oil level is reached, the float rises and operates the safety trip mechanism, which pushes the inlet needle firmly into its seating and prevents all further flow of oil.

The second function of the valve is to meter oil to the burner. This is done by a weir tube sited in the outlet of the control unit. A slotted metering stem is housed in the weir tube and variation of the height of this stem allows more or less oil to flow from the chamber to the oil burner (see Fig. 13.11). If the metering stem is depressed, the weir hole is nearly closed and only a small amount of oil is allowed to flow to the burner, and vice versa. In some instances the metering stem can be manually adjusted so that any predetermined rate of burning can be ensured, but for most part this action is controlled either directly by thermostat, or indirectly by a remote thermostat working through a heat motor sited on the control box.

Fig. 13.11. Cut-away section of typical float control, showing normal level float that operates the metering stem, and additional safety cut-out float that operates trip mechanism, forcing metering stem hard into the closed position. (BM Controls.)

Thermostatic control

The direct-acting thermostat is probably the most used device since it provides the burner and the complete appliance with a control system that functions both as a high-limit device and a secondary control so that no other form of thermostatic control is normally necessary. A vapour-pressure thermal element is employed comprising a bellows housed on the top of the control box connected to a sensing bulb, normally housed in the water circuit. The movement of the bellows is transmitted through a linkage to the metering stem of the oil-control valve.

A control of this kind is shown in Fig. 13.12. The operating bellows is carried on a platform at the top of the control box and is arranged so that it expands on a rising temperature, causing levers to be tilted. At the end of the lever system, pressure is brought to bear on the metering stem, moving this downwards and so restricting or cutting off the flow of oil. In its basic form there is a single knob, which offers a temperature selection and an off position. Various ranges are provided, depending on whether the final application is for the heating of air or hot water.

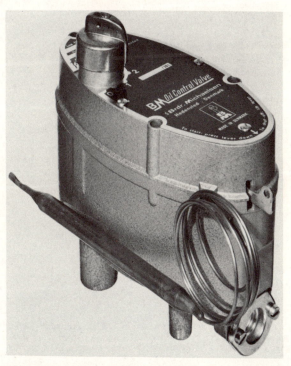

Fig. 13.12. Thermostatic version of float control, providing modulating control in relation to rises and falls of temperature at sensing probe. (BM Controls.)

To meet the increased sophistication of recent oil burners, other features can be added to the basic system to make the control meet a particular appliance need. In Fig. 13.13 a thermostatic control system is shown, which can be used in conjunction with the 'Sesto' burner shown in Fig. 13.9(c). Two sensing phials are used, one of which is sensitive to water temperature and controls the flow of oil in the normal manner as the temperature of the water rises and falls. The second sensing phial is arranged so that the burner stays on low fire whilst it is warming up, regardless of the demands from the water thermostat. In addition an extension lever will be seen protruding from the control box. This is provided to be coupled to the air intake to the burner, and leverage is so arranged that the thermostatic control mechanism provides control of the air input as well as oil input to the burner.

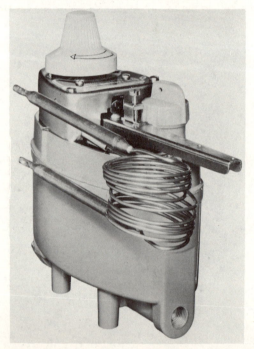

Fig. 13.13. Float control with additional thermostatic action, providing low fire control whilst the burner is warming up, and a lever to be connected to the air supply to the boiler. (BM Controls.)

The controls so far described are self-contained and require no electric or other power to assist the thermostatic function. Where oil burners are asked to work in co-operation with electric programme and time controls, the basic float control can be modified to incorporate either a solenoid, which operates the metering stem, or an electric actuator based on a heat motor. A diagrammatic sketch of a typical arrangement is shown in Fig. 13.14, where it will be seen that the heat motor, which is a strip of bi-metal around which is wrapped a heating element, is in contact with the main metering stem. On a call for heat from the thermostat, timer or programmer, the heating pad is energized and passes heat to the bi-metal strip, which warps and allows the metering stem to rise to the high fire position. On de-energizing, the bi-metal is allowed to cool and returns to its original shape, forcing the metering stem into the low or off position. Where an immediate change from high to off or high to low fire is required, a solenoid can be located on the top of the control box, exercising a similar function to that of the heat motor just described, but providing instant reaction to the availability of power.

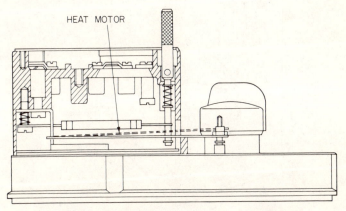

HEAT MOTOR

Fig. 13.14. Electric heat motor addition to standard float valve providing control from a remote signal. (BM Controls.)

Wall flame burners

This type of burner prepares oil for combustion in a manner inter-mediate between that of the vaporizing burner and pressure-jet atomizing burner. One such type utilizes a vertically-mounted electric motor having a hollow shaft up which oil is drawn from a well (Fig. 13.15). At the top of the shaft the oil is split into fine droplets as it is flung outwards towards the cylindrical wall of the combustion chamber. At the same time, the oil is mixed with air provided by a low-pressure fan on the top of the motor shaft. Under stabilized running conditions the mixture of air and oil strikes a ring of heated steel plates around the outer edge of the combustion zone, is completely vaporized, and burns. To start the combustion cycle an igniter is built into the burner and a switching cycle is introduced into the controls so that the igniter operates for a predetermined time when the burner is switched on.

The controls used on this type of burner vary from one type to the other. For the version described above primary control of the fuel to the burner is effected through a float chamber similar to that employed

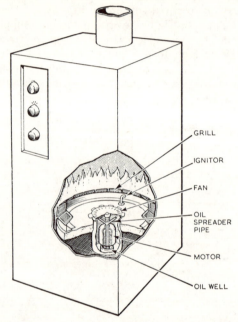

GRILL

IGNITOR

FAN

OIL
SPREADER
PIPE

MOTOR

OIL WELL

Fig. 13.15. General arrangement of a wall flame burner.

on the vaporizing burner. The height of this control is adjusted so as to maintain a certain depth of oil in the well at the base of the hollow shaft. The float control also incorporates a metering stem so that when the burner is in action the rate of consumption can be regulated from this point. The other essential controls consist of a boiler thermostat, to monitor the call for heat, and a flue thermostat to act as a safety lock-out in the event of flame failure.

The sequence of operation commences when the room thermostat or boiler thermostat calls for heat. This switches on the motor and igniter. A mixture of oil and air is then flung outwards towards the wall of the combustion zone and is ignited. As the steel plates in the outer ring rapidly heat up, the oil–air mixture striking them fully vaporizes and stabilized burning conditions are achieved. The igniter remains on for some 70 seconds, by which time the flame is fully established and the flue thermostat has received heat from the products of combustion. If heat is not registered at the flue thermostat within a predetermined time from commencement of the starting cycle, then the burner is switched off as it is assumed that the flame has not been established correctly. Similarly, if during normal running the flame is extinguished, the flue thermostat will cool and switch off the motor operating the burner.

Pressure-jet oil burners

The pressure-jet burner provides a fully-automatic means of burning oil. Many types are available with throughputs of 2·25 litre/h. The essential components in these burners (Fig. 13.16) comprise an electric motor which drives an air fan, and an oil pump. The fan is fitted with an adjustable slide so that the amount of air being driven forward into the combustion chamber can be controlled; and to the pump is added a pressure-regulating valve which controls the pressure of the oil delivered to the atomizing nozzle, and ensures that oil is only available at the nozzle when a certain minimum pressure is available. Generally there is also incorporated a strainer in the line so that blockage of the fine nozzle can be prevented. Finally, a means of ignition is provided in the form of electrodes in the path of the oil spray from the nozzle. The igniter is generally of the spark type fed through a transformer.

The operating cycle of this type of burner after the thermostat has called for heat is generally as follows. The motor is set in motion and the ignition transformer is switched on, causing the igniter to function. The fan which is attached to the motor will commence delivering air to the combustion zone and, at the same time, the oil pump will be set in

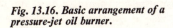
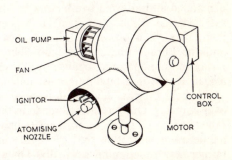

Fig. 13.16. Basic arrangement of a pressure-jet oil burner.

motion causing pressure to build up in the pressure-regulating valve. As soon as this pressure reaches a predetermined level the valve opens and allows oil under pressure to be forced through the atomizing jet. The oil spray from the nozzle then mixes with the air in the combustion zone and is ignited by the spark from the igniter.

Control of pressure-jet burners

The primary control of a pressure-jet burner is essentially not thermostatic but is concerned with ensuring that the burner operates correctly and safely. It is not concerned with the amount of heat provided by the boiler and whether this is sufficient to meet demands. The latter is the function of the secondary and high-limit controls, but as these must work in conjunction with the primary control, a general knowledge of the functions of the primary control is an advantage. Briefly, then, it must provide several things including: (1) a means of switching on the burner on a call for heat from a thermostat; (2) a means of igniting the atomized oil and air mixture as soon as this becomes available; (3) a means of detecting that a flame has been established, and be able to switch off the burner if this is not the case after a predetermined period. This safety device then continues to monitor the flame once it is established, and should it either fail completely or become weak, the control must be able to switch the burner off. Some designs allow this safety device to set in motion the complete cycle again in an attempt to re-ignite the burner after a purging period. Failure to achieve or re-establish flame on the second attempt causes the burner to lock out. It must then be manually reset.

To achieve these functions the primary control was originally a thermal device centred round a bi-metallic helix sited in the flue of the appliance and acting as a flame-temperature sensing device. The remainder of the control consisted of switches and relays with suitable timing devices so that the burner could be switched on when heat was demanded and allowed to function so long as the flue thermostat was satisfied that the correct flame had been established and was being maintained. This process continued until the call for heat was satisfied and the burner was switched off. The helix of bi-metal twisted in response to temperature changes and transmitted its turning action via a shaft to a switch sited in the instrument case in which it was housed.

The helix thermal device had invariably to be provided with change-over contacts. One circuit was made in the cold ambient condition, the other had to be made when the flue had achieved its correct running temperature indicating that the flame had been established and the burner was functioning normally. On a small drop in temperature, indicating that either the flame had been extinguished or had dropped in power, it was desirable that the thermal mechanism immediately broke its normal running circuit. This obviously could not be achieved purely by allowing the bi-metal to cool down to its ambient condition before this circuit was made, so invariably a slipping clutch mechanism was fitted to the end of the shaft. This enabled the lock-out circuit to be over-ridden as soon as the flue temperature began to rise causing the helix to twist and go to the burner-on circuit. Any further thermal movement caused by increased rises in temperature was absorbed by the slip action.

If for any reason the flue temperature began to drop the first movement of the helix would be conveyed to the switch and the circuit burner-on switch would be broken, and the lock-out circuit would be re-established.

Electrical control

Recent trends in design tend to dispense with the thermal-heat detector and use instead a photoelectric cell sensitive to the luminosity of the flame. This has been chosen because of its greater sensitivity and its instant response to flame. This is especially desirable on burners having large burning rates, for any delay in the detection of flame-

Fig. 13.17. Typical photo-resistor type oil burner control box where the complete control unit becomes a plug-in assembly on to the mounting. (Danfoss (London) Ltd.)

failure fills the combustion area and flue with a potentially dangerous
unburned oil vapour.

The photoelectric cell transmits its message to the rest of the control
circuit via an electronic amplifier circuit. Recent trends utilize the latest
type of photo-resistor which enables the amplifier to be dispensed with,
thus considerably simplifying the whole of the control box (Fig. 13.17).

Such a circuit normally operates as follows. The boiler thermostat
calls for heat and makes circuit to the control box. This switches on
the electric igniter, the motor of the pressure-jet burner, and a timing
device; and at this stage, air and atomized oil are allowed to pass into

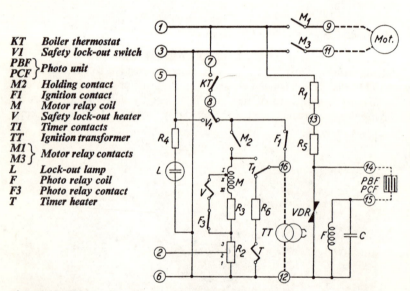

KT	Boiler thermostat
V1	Safety lock-out switch
PBF PCF	Photo unit
M2	Holding contact
F1	Ignition contact
M	Motor relay coil
V	Safety lock-out heater
T1	Timer contacts
TT	Ignition transformer
M1 M3	Motor relay contacts
L	Lock-out lamp
F	Photo relay coil
F3	Photo relay contact
T	Timer heater

Fig. 13.18. *Wiring diagram of photo-resistive type oil burner control. (Danfoss
(London) Ltd.)*

the combustion chamber. In the meantime, the timing device is approaching a point at which it will switch off all these activities. But if
flame is established it will be registered by the photoelectric cell, or
thermal sensing device, which will switch in another circuit so that the
action of the timing-device cut-out has no effect on the burner operation. If a flame is not established, the timing device takes control and
after a given number of seconds switches the burner circuit off. The
circuit of a photo-resistive oil-burner control is shown in Fig. 13.18.

High-limit and secondary controls

As already stated, it is the secondary and high-limit controls that regulate the amount of heat provided by the boiler or heat exchanger in relation to the heat demands. The high-limit thermostat is located on the appliance or boiler to which the burner is fitted. The function of this device is to ensure that regardless of the demands being made on the boiler by the secondary control the boiler is held within certain safe conditions. It normally takes the form of a thermostat housed in the waterway of the boiler or the ducting of an air heater, and is set to a certain safe-working limit. In the control circuit it is found between the secondary control and the primary control. It is normally of the re-make type, so that once high temperatures have subsided it will automatically switch on the burner and allow this to function again in the normal manner until high temperatures are predominant again.

The high-limit thermostat can use a bi-metal or vapour-pressure thermal switch with its sensing phial housed in the boiler jacket or used as a duct-mounted thermostat in the case of an air heater. Whilst they are essentially high-limit thermostats, they can, if necessary, be set well below the maximum working temperature of the appliance if the application demands a temperature lower than this. For instance, a high-limit thermostat used on an oil-fired boiler supplying central-heating requirements could be set down to 65°C in mild weather, even though the maximum safe-working limit would be in the region of 93°C.

Some circuits employ two thermostats, one as a normal regulating device providing adjustable control over the range of temperature required, whilst a second thermostat acts as a true high-limit thermostat set to the safe-working temperature of the boiler. This then acts as a monitor on the adjustable thermostat and gives a second line of defence if the first one should fail. This is an especially desirable feature when the ordinary adjustable thermostat is not of the fail-safe type, and when, should the thermal system fail, the contacts will be set permanently on. For this reason most high-limit thermostats are of the fail-safe variety, so that should anything happen to the thermal system during the course of their life, the contacts will be sent to the fully-open position.

The secondary control on such an application can take the form of room thermostats, ducting-discharge thermostats, automatic timing devices, humidity controls, pressure controls (Fig. 13.19). All these will demand heat so long as the condition they are controlling is below set-point and so long as the timing devices, etc., also demand that the appliance should be working. They will be completely insensitive to

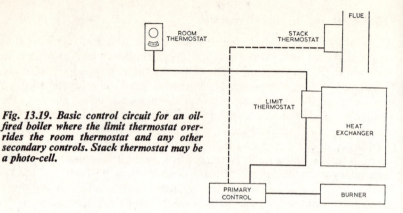

Fig. 13.19. Basic control circuit for an oil-fired boiler where the limit thermostat overrides the room thermostat and any other secondary controls. Stack thermostat may be a photo-cell.

conditions both in the boiler and in the burner, so that their demands for heat will only be met so long as the high-limit control is satisfied that conditions in the boiler are normal and the primary controls are satisfied that the burner is functioning in a satisfactory manner.

High-limit and secondary controls for vaporizing burners and wall flame burners

The primary control of the vaporizing and wall flame burner comprises the oil-level device which is an integral part of the burner assembly. This control meters oil to the appliance and ensures that the burner is functioning correctly. In some instances where only limited control is required a thermal device is built into the control valve and this satisfactorily controls the output of the burner in relation to application demands. However, for boilers fitted to central-heating or air-heating circuits it would be impossible to control the heat output from the appliance adequately purely by built-in controls and there will be a demand for remote control.

When this type of system is employed there is an additional need for a high-limit control to be fitted to the appliance to protect it from excess demands from the secondary controls sited remotely. In addition when secondary controls call for intermittent heat, the burner must be made fully automatic, and this in turn calls for additional devices in the primary-control circuit.

When remote or automatic control is considered for vaporizing burners, it is essential that the burner operates safely under all conditions and demands. This creates a need for a sequence controller, a typical example of which is shown in Fig. 13.20. This is a series of cams driven by a synchronous motor, which provides a lighting-up sequence,

Fig. 13.20. Cam-operated sequence controller. (Redfyre Ltd.)

which, once initiated, will ensure that a safe sequence of operation occurs. Once the boiler has been switched on, the lighting up, running and shut-down sequence is controlled by the automatic unit. The start-up and lighting sequence occupies some 2 min and the shut-down sequence 5 min. On switching the boiler on, the electric ignition, oil supply and fan are all energized in a set sequence, and having established a flame in the burner the electric ignition device is then automatically switched off. The boiler now continues to operate at the required temperature, until the boiler thermostat is satisfied and breaks the circuit. The automatic control box then switches off the oil supply and allows the fan to run for 5 min to purge the burner before this also is automatically switched off. If during this purge period the boiler thermostat contact should remake, either through a drop in the temperature of the water or by the thermostat knob being turned to a higher setting, the boiler cannot relight until the purge period is completed, when the full lighting up sequence will be initiated.

DOMESTIC COOKING CONTROLS

In the manufacture of domestic and commercial cooking equipment those appliances using gas or electricity take pride of place, and for the most part thermostatic control of the oven at least is offered as a standard part of the equipment. In this respect the thermostat serves a very useful purpose in that regardless of the source of heat the cook can set the range knob to a predetermined position and obtain the cooking results that she requires without having a detailed knowledge of the differing heating problems created by the use of either gas or electricity. To the control designer, however, dependent on whether he is attempting to control gas or electricity, the problems are quite different. The gas thermostat, or gas control, can act as an infinitely variable device and can proportion or modulate the flow of gas in relation to temperature changes in the oven. Electrical energy can only be controlled on a stop-start basis and, therefore, thermostats controlling it tend to assimulate on-off or two-position control, with the accompanying problems of overshoot, although this can be minimized by the use of fine differential, or creep action, switching. Nevertheless, this fundamental difference between the methods of control remains and in the following chapter some information is provided as to how basic principles are applied to the actual practical problems of designing and applying thermostatic control to domestic cooking equipment.

GAS COOKER CONTROLS

The domestic gas cooker was one of the first domestic appliances to be equipped with automatic control, and it is still in the forefront of automatic control in the domestic appliance industry. The first thermostatically-controlled ovens in quantity production appeared in the early 1920s and soon came into general favour. More recently, thermostatic control of the hotplate burner has become popular, offering full temperature control over all the cooking operations that are possible on the top burner.

Oven temperature control

The control of temperature in the oven is defined in British Standard 1250; and the appliance is expected to produce any temperature within the cooking range down to approximately 116°C within fine limits.

The heat for the oven is generally obtained from an aerated gas burner at the bottom of the back of the oven or from two such burners, one on either side. The hot products of combustion are made to circulate in the oven space and are then vented to atmosphere. The venting can vary from one design to another and is an important factor in determining the arrangement of the thermostatic control. Two types of oven are shown in Fig. 14.1. In one type the combustion products are vented out of the top of the oven and in the other a bottom outlet flue

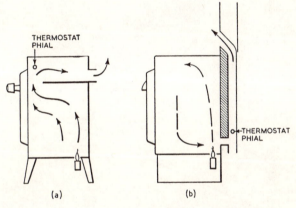

Fig. 14.1. Thermostatic phial positioning in two types of domestic gas cooker.

is provided to evacuate the products of combustion up the back of the cooker. The requirement of an effective thermostatic control is that it shall ensure that the temperature of the centre of the oven is maintained at a constant temperature for in most cooking operations this is where the food is sited. The thermostat therefore has to be placed in a position where it is out of the way of the cooking and cleaning operations, but where it can sense a temperature which bears a definite relationship to the temperature existing in the centre of the oven.

Each oven design will have a particular temperature gradient between the centre of the oven and the position of the sensing phial of the thermostat. This gradient depends on the pattern of the flow of the products of combustion and the siting of the thermostat. This means that the degree of offset or the difference between the oven-centre temperature and the actual temperature being sensed by the thermostat phial will vary considerably from one cooker design to another. A typical relationship between these two temperatures is shown in Fig. 14.2. The cooker manufacturer decides upon the centre-oven temperature required for the various cooking operations, and then arranges for each of these temperatures to occur when the thermostat knob is turned to the appropriate setting. In the United States and, to some extent, in Europe, the thermostat range knobs are marked in actual temperatures, but in the United Kingdom the accepted standard marking is in numbers from $\frac{1}{4}$, $\frac{1}{2}$, 1, etc., up to 9.

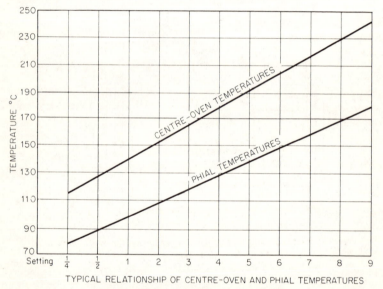

TYPICAL RELATIONSHIP OF CENTRE-OVEN AND PHIAL TEMPERATURES

Fig. 14.2. Graph of typical thermostat offset.

The position of the thermostat is also important as this not only ensures the correct cooking temperature, but also to some extent governs its sensitivity. Hot gases are poor conductors of heat; moreover, the thermostat sensing element may have a considerable thermal mass of its own. If the necessary precautions are not taken, these two factors could produce considerable time lags in reaction to temperature, and cause hunting on either side of the control point and slow reaction to changed conditions. However, experience has shown how the position of the thermostat element should be chosen to utilize the flow of hot gases to the best advantage; and in oven design today a high degree of sensitivity is achieved.

Rod-and-tube thermostats

In the U.K. for many years the rod-and-tube thermostat was used exclusively, because of its simplicity and reliability. The differential expansion of a brass tube and Invar rod introduced few problems of range or calibration stability, whilst the all-metal construction of this type made it very robust—a distinct advantage when, in the days of its introduction, fine mechanisms on domestic appliances were the exception rather than the rule. This design is still in use on a large number of cookers and, whilst considerable differences in external detail are found, the basic internal design is generally the same. The principles of operation of the thermal element, consisting of the brass tube and the central Invar rod, have been explained in Chapter 1, where an outline description of its application to gas-cooker controls was given. It is sufficient here to recall that when the temperature rises the rod is withdrawn from the thermostat body, allowing the spring-loaded valve to make contact with its seat; this cuts off the gas except for a small by-pass flow that keeps the burner alight. For temperature selection, the position of the valve in relation to its seat is manually adjusted by turning the range knob. For example, if a high oven temperature is required, it is arranged that the gas supply is not cut down until the brass tube has expanded considerably.

The choice of position for the rod-and-tube thermostat is to some degree limited. Being a self-contained unit with the range adjustment fixed at the end of the thermal element, it has to be sited so that the thermal element is in a good position for sensing temperatures, and yet where the range adjustment, which will be on the exterior of the cooker, is easily accessible and fits in with the general styling. Some compromise is often necessary between these factors; and, with increased emphasis placed on styling, more designers are employing the liquid-expansion thermostat.

Liquid-expansion thermostats

This type, with its sensing bulb connected to the thermostatic valve by flexible capillary, offers considerable latitude in the placing of the thermostat and the sensing phial. The thermostatic valve can be placed in a position that suits the styling of the cooker, whilst the thermal element can be located in the most convenient position in the

oven, both for sensing of the temperature and keeping it clear from
cooking and cleaning operations. In some instances it is taken right
out of the oven and put in the flue outlet, leaving the oven clear of any
obstruction. The mechanical detail of the liquid oven thermostat can
vary considerably from one design to the other; two typical types are
those operated by diaphragm and by bellows.

Diaphragm type

A typical example of this application is shown in Fig. 14.3. In this
model the thermostat has an on-off tap. This, and the temperature
setting cam, are controlled by a single knob. In this way the need for a
separate oven tap, knob, and associate pipework, is eliminated. The use
of a cam enables non-linear temperature settings to be achieved with
equally spaced knob markings, and by changing the cam profile
varying degrees of non-linearity can be provided to suit different oven
requirements.

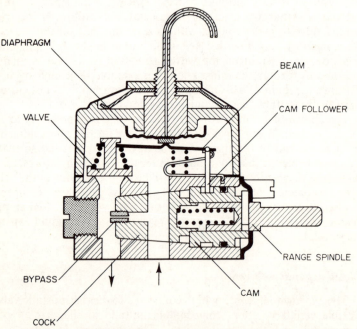

DIAPHRAGM

BEAM

CAM FOLLOWER

VALVE

RANGE SPINDLE

BYPASS

CAM

COCK

*Fig. 14.3. General arrangement of a liquid oven thermostat with
diaphragm-type capsule. Valve is shown in the closed position.*
(Concentric.)

The operation of this thermostat can be seen from the diagrammatic sketch. The knob is turned to a predetermined setting number and during the first part of the turning action the tap is turned so that gas is allowed to pass through the thermostat. From then onwards the tap is maintained in the 'on' position by a gallery running round it, so that further turning through the thermostat mark numbers does not have an effect on the flow of gas. Rotation of the knob will also cause the cam to rotate, and in so doing raise or lower the cam follower and cause this to tilt the operating lever and thus predetermine the initial position of the variable valve. Once the oven is lit the hydraulic expansion of the liquid in the sensing bulb is transmitted through the capillary to the diaphragm, the centre of which is in contact with the operating lever. Thus, as the diaphragm receives the expanding liquid and moves forward, it depresses the lever and in so doing tends to close the variable valve. The by-pass gas rate is maintained through a controlled orifice in the bottom of the plug.

Most thermostats provide a means of modifying the calibration on site. In the case of the thermostat shown in Fig. 14.3 this is achieved by loosening the locking nut on the back of the diaphragm stud and by means of a tommy bar rotating the screwed boss housing the diaphragm stud, causing the diaphragm to move farther in or out of the body of the thermostat, and thus resetting the initial relationship with the operating lever. Having achieved the desired alteration, the locknut is tightened again and the thermostat checked for the effect of the adjustment being made.

Bellows type

In this case (Fig. 14.4) the sensing phial is part of a hydraulic thermal system and is connected to fine bore capillary, which in turn is connected to a bellows system in the head of the gas valve assembly. The bellows is in contact with a valve spindle, which pushes a valve towards a seat thus throttling gas as it passes through the tap in the body. Temperature selection is achieved by rotating an outer sleeve on the tap spindle assembly, and taking with it a cam. This in turn operates a push-rod, which alters the position of a hinged beam against which the thermal bellows assembly is pressed; thus rotation of the sleeve and cam causes these other members in the lever system to pre-position the variable valve and therefore to control the amount of movement required of it before it throttles the gas supply to the burner. The main gas supply to the burner passes through the plug in the body and through the variable valve, and then by way of a side channel (not shown in the drawing) to the main outlet of the thermostat body. In addition to this passageway, a by-pass rate is provided by means of a jet housed in the base of the plug. This will pass gas, regardless of whether the variable valve being controlled by the thermostat system is in the open or closed position.

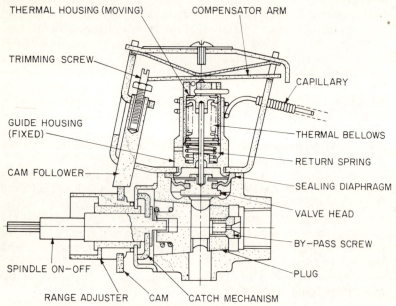

Fig. 14.4. *Thermostatic gas valve using bellows type liquid thermal system. (Teddington Autocontrols Ltd.)*

The gas tap can be operated independently of the thermostat range dial, its function being to admit gas to the burner in the first instance and if necessary pre-set the gas rate so that over-shoot on very light loads at low temperature is avoided, a feature which becomes important when this thermostat is applied to the thermostatic control of hotplate burners described later in this chapter.

Oven thermostatic-control design

In designing a thermostatic control for a cooker or other small gas-heated ovens certain factors must be known about the oven performance itself. In the first place the temperature over which thermostatic control is desired must be established. These are the temperatures at the centre of the oven space and in all oven design work are best measured by a thermocouple in a blackened ball.

Next, a decision has to be made as to where the thermostatic phial is to be located and the temperature difference between the centre of oven and the phial location must be established. This difference is not necessarily constant over the whole of the operating range; it is therefore desirable that the phial temperature and centre-oven temperature relationship be measured over the complete operating temperature range. Furthermore, as both the rod-and-tube thermal element and the hydraulic thermal element sense temperatures over the length of their sensitive portion they are registering average temperatures over that length rather than spot temperatures at any one particular point. Therefore, it is not entirely satisfactory to assess phial temperatures by thermocouple reading or other spot measurement. It is preferable in early experiments to assess the hydraulic or thermal movement of the sensing element when sited in its correct position in the oven, noting the movement obtained against a range of centre-oven temperatures. The thermal element can then be removed from the oven and placed in an oil bath, the temperature of which is gradually increased so that as the thermal element responds and produces movement the temperature of the oil bath can be noted against the same movement obtained when the element was in the oven. In this way a more accurate picture of actual phial temperatures against centre-oven temperature can be obtained.

When the temperature range over which the thermostat is to operate, and to be finally calibrated, has been established, it is necessary to define the length of thermal rod or the volume of hydraulic sensing phial to produce sufficient movement to bring the valve down to its seat at the correct temperature, and yet provide sufficient movement so that reasonable gas rates can be maintained right up to the point where the thermostat is set to maintain that temperature. The degree of thermal movement required therefore depends to some degree on the diameter of the gas valve employed. A small valve would need to be held off its seat right to the point of temperature control if reasonably quick heating-up rates were to be obtained; therefore, a large amount of thermal movement would be desirable so that only in the last few degrees of temperature rise to the control point would the valve begin to throttle the gas supply. On the other hand, a large valve would maintain high gas rates almost to the point of closure; therefore such a valve would only require a small thermal movement per degree to enable it to maintain reasonable gas rates until the point of control was reached. Experiments on valve size and thermal activity will therefore decide both the heating-up rate and the ability to stabilize quickly, and also the response rate in relation to load changes and temperature drops caused by such incidents as opening the oven door and placing cold cooking loads into the oven.

Finally, when the thermal movement necessary and the temperature range over which this will be provided have been fixed, it will be necessary to decide on the range mechanism to be employed so that the thermostat can be set to any point in its declared range. This generally resolves itself into a threaded screw or a cam terminating in a range knob marked suitably in terms of degrees of temperature or cooking-mark numbers. The rotation of this knob is normally limited to rather less than 360° angular rotation and is arranged to preposition the valve away from its seat. Mechanical linkage between the valve and the thermal element then allows the valve to be brought closer to the seat and finally on to it as thermal movement develops in relation to rise in temperature. With a knowledge of the full amount of thermal movement that will be produced by the sensing element, it is comparatively simple to arrange for the valve to be displaced off its seat by this total amount of movement by rotation on the range screw. It will obviously be necessary to use a thread on the range screw of a pitch which will lift the valve off its seat by a distance equal to the full thermal movement available for a rotation of the screw through 360° or any percentage of a full circle desired by the appliance manufacturer.

Automatic time programming

Trends in gas–cooking appliances suggest that in future there will be a departure from the simple schemes described above as the gas appliances become more complex. There are already in existence a number of appliances offering automatic time programming. The user has only to place the food in the oven and to set a time switch and thermostat. Thereafter the oven can be left unattended: it automatically switches itself on at a certain hour, cooks the food, and then switches itself off. With the increased use of frozen and pre-cooked food these automatic cycles can become more complex. Already in the United States appliances are available which offer a programme whereby food is first de-frozen, then cooked, and subsequently allowed to cool and is held at a temperature which keeps it hot enough for eating but not so hot that dehydration takes place. These trends call for more complicated thermo-static control systems, together with flame-failure and timing devices.

One of the main problems is that in the 'keep warm' cycle temperatures lower than those normally required in cooking operations are necessary, with the consequent need for very low gas rates. A burner capable of supplying heat to maintain a temperature of 290°C can be hard put to maintain flame stability if the gas throughput is reduced to maintain temperatures of the order of 50°–60°C. Slight draughts or rapid door closures might extinguish the flame and cause unburned gas to escape.

If such a burner is required to work satisfactorily on this type of work there is a possibility that thermostatic gas control will be required to provide on-off control rather than modulating control. The oven temperature is then maintained at a reasonably constant figure by a

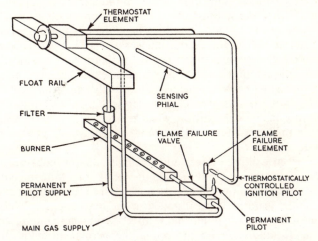

Fig. 14.5. General arrangement for on-off type thermostatic control on a domestic gas cooker.

series of bursts of full heat at full gas rate. This arrangement requires a fool-proof automatic re-ignition system to ignite the burner every time the gas is switched on by the thermostat, and effective flame-failure arrangements must be provided to protect this arrangement.

Schemes providing such control are already available in America and undoubtedly it will only be a matter of time before they appear on English cookers. The final form that such servo-mechanism gas valves and thermostats are likely to take is difficult to predict, but a basic scheme generally as described below could be typical of the systems employed. Fig. 14.5 shows the general layout of the scheme, in which it will be seen that the thermostat is sited at the top of the cooker, and the sensing phial from the thermostat is in a normal position in the oven. The mains gas flows from the thermostat or alternatively from the float rail to the burner. However, before it enters the burner, it has to pass through a flame-failure valve which is controlled by another thermal sensing element. This element normally closes the valve, which is only open to allow gas to flow to the burner when the sensing element protruding from the valve assembly is heated. It will be noticed that a small pilot supply is taken from the float rail and feeds a permanent pilot which remains on during normal cooker usage. A second pilot gas supply comes from the thermostat to feed an ignition pilot. When the thermostat is calling for heat, gas flows through to this ignition pilot, which is ignited by the permanent pilot. The long ignition pilot flame passes across the sensing unit of the automatic flame-failure cut-off valve and, when the sensing unit is heated, it opens the valve allowing gas to flow to the burner. The pilot flame ignites the burner and the oven heats up. When the thermostat is satisfied, i.e. when the heat of the oven reaches the prescribed point, the thermostat cuts off the flow of gas to the ignition pilot, which is then extinguished. The sensing element of the flame-failure safety control valve cools and cuts off the flow of gas to the burner. The burner will then stay out until the temperature in the oven drops, when the thermostat valve will again allow gas to flow to the ignition pilot and the whole process is then repeated.

Where electric power is already available on the cooker, electric timers can be introduced and provide the facility of automatic cooking. A typical scheme is shown in Fig. 14.6 where the electric timer controls a heat motor operator valve installed in the oven supply.

Thermostatic control is normally direct acting modulating control with a flame failure device protecting the permanent pilot. In this instance the electric valve is normally open and the timer energizes it in the closed position during the delay part of the automatic cycle. On manual operation or in the event of power failure, the electric valve is open and the cooker can be used in the normal manner.

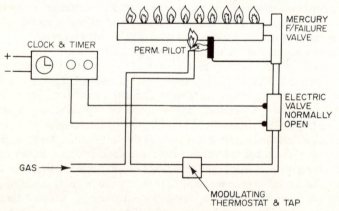

Fig. 14.6. Automatic oven control circuit using electric timer and normally open electric valve.

Gas hotplate controls

Following the general trend to make all cooking operations auto-matic, the modern gas cooker sometimes incorporates a thermostat in at least one of the top plate burners so that all cooking processes normally carried out by the burners so fitted can be controlled automatically. The thermostatic control employed (Fig. 14.7) is a natural development of the hydraulic-expansion thermostat already described as a typical gas-oven control, i.e. a thermal system comprising a sensing phial attached to the main control valve by a length of capillary and trans-mitting movement to a gas valve by means of diaphragm or bellows movement. In the hotplate control, the sensing phial normally takes the form of a disc sited in the middle of the burner. It is generally arranged that the free-standing position of the disc is slightly proud of the pan supports so that when a vessel is in position it comes in hard contact with the sensing disc and depresses it until the disc and pan supports are at the same level, leaving the disc in firm contact with the base of the vessel. To achieve this flexibility the thermal element is generally supported in a sleeve, obtaining its spring action by virtue of the capillary being coiled behind it and additional springs.

In this application it is desirable that the sensing disc receives as much heat as possible from the base of the pan and that it is influenced as little as possible by extraneous sources of heat such as radiant heat from the gas flame and the mass of the burner casting, and conducted heat from the surrounding cooker parts. To assist in this respect, the outer shield is generally constructed of stainless steel, which has poor heat-conducting qualities and can also reflect the radiant heat that would otherwise be absorbed by the sensing phial from the burner

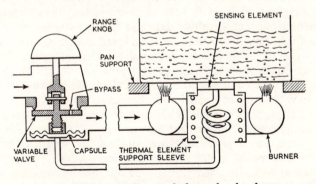

Fig. 14.7. Thermostatic control of a gas hotplate burner.

flame. In most models it is also arranged that as the sensing phial is depressed it sinks within the outer shield which is separately sprung. The centre hole in the burner is sized so that in addition to accommodating the support sleeve for the sensing phial there is extra air space to enable some ventilation to act as a cooling influence on the support sleeve, and also to provide additional air for complete combustion in the burner itself.

The range-control mechanism is similar to that employed on the liquid oven thermostat. Adjustment of a range knob alters the position of the thermostatic gas-supply valve in relation to its seat. The burner is prevented from being completely extinguished by a by-pass of gas which is available if the thermostatic valve is shut completely by a rise in temperature.

Design of hotplate thermostatic controls

Whilst this basic control system is comparatively simple, the wide range of cooking temperatures required, and the widely different load factors created by various cooking loads, makes the final selection of burner design, heat input, thermal sensitivity, and throttling control extremely complex.

Consider first the diversity of loads likely to call for automatic control. At one extreme, the smallest load may be a cup of milk in a small saucepan which is to be heated up as quickly as possible to near boiling point and held there without boiling over. At the other extreme, deep-fat frying operations can demand that up to 2 kg of fat is heated as quickly as possible to a temperature in the region of 200°C. The control for the small amount of milk could be accomplished quite easily if the rate of heat input was restricted, or an early anticipation of heat was registered within the thermal device so that any overshoot was precluded. However, these two desirable features cannot be tolerated when both the same burner and the same thermostatic control are called upon to meet the upper extreme condition for deep-fat frying, where any restriction on the heat input, or any device incorporating early anticipation of heat rise, would unacceptably prolong the cooking operation and reduce the rate of response to changed conditions.

The other problem that has to be considered is that associated with all proportional controls, in that for a given temperature set-point on the control knob and a given phial temperature, the thermal valve will assume a fixed position with a given heat throughput. Thus, when set to regulate at 90°C, the valve will assume this position regardless of the fact that it may be expected to hold a cup of milk or anything up to

4·5 litres of water at this temperature. Obviously for the same heat throughput it is unable to provide the same final temperature to the two amounts of liquid; therefore according to the throttling curve of the particular valve, a definite offset will occur.

The final solution to these problems would be found in a sensing phial with zero lag, transmitting on-off control to the gas burner, in which case all the variations on the above condition could be met. However, practical considerations rule such an arrangement out, and the final solution is in fact a compromise of these factors. The burner heat input is normally fixed at some 10·5 MJ input, thus providing a reasonably quick heat input for the heavier load conditions, and yet still providing reasonable efficiency when vessels having small bases are used. The proportional band of the thermostat is governed by the movement per degree of the thermostatic valve, which, in turn, is a function of the hydraulic expansion and the size of the valve itself. These two limiting factors have produced a proportional band of the order of some 0·03 m³ for 1°C. Thus a typical arrangement would be for the valve to move from the fully-open position to the by-pass position for a change of some 22°C.

Sensitivity

The final factor, that of sensitivity, or temperature lag between the temperature of the liquid in the pan and the temperature of the liquid in the thermal head, is one over which little control can be effected. No one as yet is able to dictate to the housewife what type of cooking vessel she will use, and how thick its base will be, or of what metal it will be made. Therefore, this considerable variable has to be accepted as an uncalculable factor, and the manufacturer is left with the problem of providing a thermal head which must provide the minimum lag in conveying temperatures received by its outer casing to the sensing liquid inside and ensuring that extraneous heat from the burner and surrounding castings is kept as low as possible.

Fig. 14.8 shows two typical curves obtained on a commercially-available pan control, the lowest curve showing the type of control available on the low heat input problem where 0·5 litre of water has to be raised to a point near 82°C and held there indefinitely. The other curve shows the condition that applies at the other end of the cooking range, where deep-fat frying takes place and where it is necessary to raise a quantity of fat or oil as quickly as possible to a temperature in the region of 200°C and hold it there for the length of the cooking process.

Finally, the burner manufacturer has a considerable problem in that, if the combined thermostat and burner is to maintain small amounts of liquid at low temperatures, such as simmering at 82°C, the burner must be capable of maintaining flame stability at extremely low gas rates. In America, where burners tend to be larger and have higher gas inputs, this problem has been even more acute, to a point where instead of using proportional control only, extra devices have been incorporated so that below a certain gas rate the valve snaps to the fully-closed position leaving a permanent pilot burning so that when the valve again snaps open the flame is re-established. Also the use of large burners has meant that it is virtually impossible to control the temperature of small amounts of liquid starting with the burner on full heat.

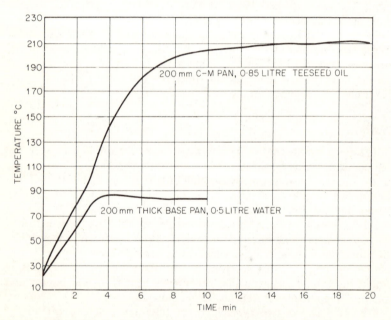

Fig. 14.8. Typical control curves for a gas hotplate thermostat.

The rate of temperature rise of these small amounts can be so rapid that the sensing phial is unable to keep up with it and temperature overshoot becomes unavoidable. Such a condition in the case of small amounts of milk results in boiling over, with the risk of extinguishing the burner.

It has therefore been found necessary to incorporate a means of throttling the gas rate by manual adjustment, so that under these conditions the initial gas rate is retarded and the heat input can be so regulated that the contents in the vessel only heat up at a rate with which the sensing element can keep pace. Most English designs also incorporate a means of manually controlling the initial gas rate, although experience has shown that so long as the maximum throughput of the burner is restricted to some 10·5 MJ a good hotplate control should be able to offer sufficiently sensitive control to dispense with this safeguard.

Fig. 14.8 shows typical curves obtainable on controls already available; although these controls are commercially reproducible in large quantities and find acceptance on most cookers there is still much work to be done to improve them. As was stated earlier, the optimum performance could be obtained from a sensing device having zero temperature lag, coupled to a valve with on-off control, since such a combination has the ability to overcome rapid rises in temperature and variations in load. In America there is already available an electrically-operated arrangement where the gas supply is controlled through a solenoid-operated valve, so that the gas is either fully on or off; and it is possible that a variation of this arrangement will be the final solution to gas hotplate control. This, however, involves the obvious complication of applying mains electricity voltage to gas cookers, and it is probable that the ultimate solution of the problem lies in the use of extremely sensitive thermal elements operating proportional valves with very fine throttling curves.

ELECTRIC COOKER CONTROLS

Early electric-cooker ovens consisted essentially of a metal box with firebrick elements housed in the top and bottom of the oven. The firebrick elements consisted of heating coils embedded in ceramic, and produced a combination of radiant and convection heat for the cooking process. During this period thermostatic control was not generally available and the majority of ovens were provided with a three-heat type of switch using the series-parallel circuit. This type of design was followed by one with side bricks enclosed in stainless steel. This increased the radiant surface and reduced its temperature, and so provided more convection heat, a trend which was assisted by using the pressed metal shelf racks on the sides of the cooker as a convection channel. Thermostatic control began with the rod-and-tube thermostat housed in the top of the oven space. This sensed the temperature of the rising convected air and controlled the oven by means of a switch operated by the expansion of the brass tube.

Modern electric cookers use the sheath-type heating element mounted in the side walls of the oven; in many types this is extended beneath the bottom of the oven to provide a small percentage of bottom heat. This latter feature is employed to counteract the natural stratification which can occur by relying purely on the convection currents of air as they pass across the elements in the side of the oven and circulate in the top area. The basic arrangement is shown in Fig. 14.9.

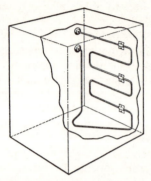

Fig. 14.9. Diagrammatic arrangement showing position of the heating element in a modern electric oven.

Whilst cooker styling remained comparatively simple the rod-and-tube type switch could be successfully applied, and it was not difficult to arrange for an extension spindle from the switch to the front of the cooker to enable a knob marked in degrees of temperature to be sited on the side of the cooker. However, modern styling made the application of rods and linkages increasingly difficult; and now, for reasons already stated in connexion with gas-cooker controls, a large number of cookers employ hydraulically-operated thermostats.

Rod-and-tube thermostats

Electric cooker thermostats of the rod-and-tube type have been widely used for many years. In principle, they differ from those used for gas cookers only in that they operate electric switches instead of gas valves. A description of a typical thermostat of this kind was given in Chapter 1. Range adjustment is effected by an adjusting screw which, via the pivoted link, changes the initial position of the moving contact arm—see Fig. 1.3—and so changes the gap between the fixed contact and moving contact for any given temperature sensed by the rod and tube. A degree of snap-action is effected over a predetermined differential.

Liquid-expansion thermostats

Fig. 14.10 shows a typical layout for the fine-differential type of thermostat frequently used on the domestic electric cooker and other small electrically-heated ovens. This type of thermostat uses the hydraulic liquid-expansion system where thermal expansion of a liquid in a sensing phial is conveyed to a diaphragm via capillary tubing. The diaphragm is sited in a switch housing and forms an integral part of the transmission circuit. Referring to the diagrammatic layout shown in Fig. 14.10, and viewing this generally from left to right, spindle *A* is normally fitted with a range knob, the rotation of which pre-sets the temperature at which the switch is to break circuit. The spindle has an

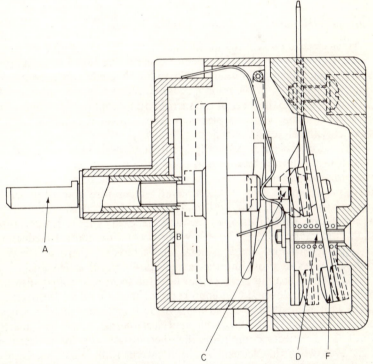

Fig. 14.10. Cross-section of an electric cooker control using hydraulic thermal movement. (Diamond H Switches Ltd.)

internal thread which supports a mounting plate *B* for the diaphragm unit. Rotation of the spindle varies the height to which the diaphragm unit is mounted in relation to the rest of the switch assembly. The diaphragm unit is spring-loaded so that at all times it is in firm contact with the range spindle *A*. Hydraulic expansion, conveyed to the diaphragm via the sensing phial, causes it to expand and thus move its centre stop *C* forward. This movement is transmitted to the moving contact arm by way of an insulating button, and contact *F* thereby open circuits the load. The calibration screw *D* is mounted in the fixed contact arm and is used for factory pre-setting purposes. In addition, it may be used to make small adjustments on installation. The whole system is carefully balanced so that contact movement in relation to thermal movement provides accurate control and fine differentials on temperatures in the cooker, and yet at the same time provides sufficient contact movement to prevent unacceptable arcing or high rates of contact wear.

Thermostatic control of electric hotplate elements

Most electric cooker manufacturers now offer thermostatic control on at least one of the hotplate elements as an alternative to the normal energy-regulator type of control. These hotplate controls usually have a sensing pad sited in the middle of the hotplate element and normally protruding above the level of the surrounding detail so that when a pan or other cooking vessel is placed on the element it comes into hard contact with the sensing head.

The sensing head is generally spring-loaded so that as the pan is placed in position it recedes and allows the pan to come into contact with the electric heating element. A control knob is provided on the facia panel which gives a selection of temperature control generally ranging from 60°C at the bottom, to control such cooking operations as simmering, whilst the top of the range generally rises to temperatures slightly in excess of 200°C to provide temperature control of deep-fat frying. Apart from these general characteristics there are several different ways of providing the temperature control called for.

The electric heating element can take the form of a coiled sheathed element which glows red on energizing, or in the modern form a solid plate with elements embedded in it. In both instances there is a basic control problem in that the element has thermal mass and inertia. Both these aspects complicate the thermostatic-control problem in that, even when the element has been switched off, it still has heat to discharge which will continue to produce a rise in temperature in the contents of the cooking vessel. Any thermostatic control device, therefore, must *anticipate* temperature rise and either cut off or taper off the power input to the element as the set-point temperature is approached. Each type of heater element needs different treatment in this respect, according to its thermal mass and its wattage input. Failure to observe this fundamental feature can cause excessive overshooting as the control point is reached, which would be objectionable to the housewife particularly when boiling milk.

Design considerations

Often when faced with the problem of fixing the *anticipation point* the manufacturer is confronted with two conflicting requirements. In the first place, he must provide considerable anticipation so that when a small amount of liquid is being heated at a low temperature the anticipation is sufficient to prevent overshoot. On the other hand the same thermal device has to control large amounts of liquid fat at 200°C and, here, early anticipation would cause the temperature rise at this end of the range to be very slow.

The design of the sensing head is equally important, and the fact that several entirely different designs exist shows that there is no easy answer to this problem. The basic requirements are that it should quickly sense changes of temperature within the pan, and that it should be as insensitive as possible to extraneous sources of heat such as radiation from the element and conduction from surrounding metal parts.

Finally, there is the problem that most of these devices are proportioning in their basic principle, and tend to suffer from the limitation of this type of action in that the final stabilizing point can be off-set from the temperature set-point due to changes in load. To take an extreme example, a controller could be set to provide slow-boil conditions for a cup of water with the dial set to 100°C and under these conditions hold the liquid at a slow boil temperature. If, however, the amount of liquid is suddenly increased to several pints, then the control temperature will settle at a point somewhere below boiling point. This problem has been overcome to some extent by providing the user with a selection of set-points for the same temperature at varying load conditions. Much work is being carried on to overcome this problem with the use of proportional-plus-integral control devices.

Control arrangements

Two types of sensing element are in general use, one using liquid-expansion, the other using electrical energy and relying on an electrical resistance thermal element for registering temperature changes. As was stated earlier, it is not sufficient to provide on-off control to the whole of the power element; some degree of temperature anticipation has to be provided. In its simplest form this can take the form of multi-step switching. The electric element is divided into two sections each of which is controlled through one stage of the thermostat switching. Thus, as the temperature of the vessel rises, the hydraulic-expansion in the element at first switches off one section of the element, leaving the remaining energized section of the element to carry the temperature of the vessel up to the set-point. In this way excessive overshoot is avoided and a considerable degree of control is procured.

Other arrangements use the heat-motor principle as a means of introducing anticipation and tapering down the power input as the temperature of the vessel approaches its set-point. Here, full power is available until the temperature of the vessel reaches a predetermined distance from the set-point. At this stage a bi-metallic heat motor is brought into play which begins to reduce the heat input to a series of on-off cycles, and so reduces the total heat input. This is achieved by having a strip of bi-metal which is a part of the switch mechanism moving the contact. The bi-metal receives heat either from the element load which is passed through it, or from a separate heater wound round it. In a typical arrangement it is so arranged that heat is imparted to the bi-metal only when the main contacts are closed and the power is

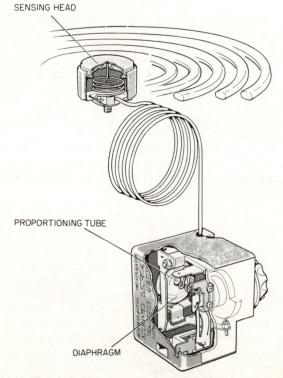

Fig. 14.11. Electric hotplate thermostatic control. (G.E.C. SUNVIC.)

available to the element. Thus during this period the bi-metal is warmed up and distorts to a point where the main contacts open and cut off power to the heater element. As soon as the circuit is broken the bi-metal begins to cool and returns to its former shape, and in so doing makes contact again for the power to the element. This cycle is repeated continuously. Some systems arrange for the cycle rate to be at a fixed timing once the anticipation point has been reached, whilst in other systems the cycle time varies and diminishes as the control point is reached.

One of the most popular forms of electric hotplate control in current use is shown in Fig. 14.11. This employs a hydraulic thermal system, the sensing component taking the part of a pad engaging the pan

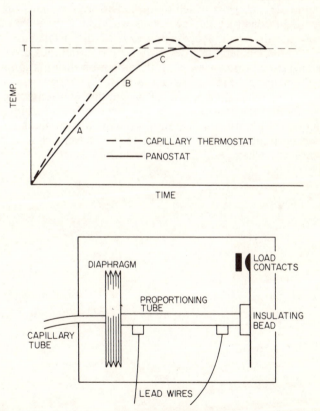

Fig. 14.12. Anticipator arrangement electric hotplate control. (G.E.C. SUNVIC.)

bottom. In this instance the anticipatory action necessary for temperature control on light loads makes use of a 'proportioning tube'. Fig. 14.12 illustrates how this is achieved. The dotted line shows a typical curve for a capillary thermostat set at a temperature T. The overshoot temperature is unsatisfactory in a domestic cooker hotplate application where milk could boil over under this condition. It is therefore necessary to provide a heating curve as shown by the solid line with no overshoot. This is done by passing the hotplate load current through the proportioning tube which transmits the mechanical expansion of the diaphragm to the switch.

The proportioning tube expands longitudinally due to the heat generated by the flow of the current, and at a position nearing A on the curve. This expansion, together with the diaphragm expansion from the sensing head, causes the switch to open. The proportioning tube, being of very light construction, soon cools when the current flow is interrupted, and allows the load switch to close again after a very short period. As the temperature rises further short switch-offs occur, but as the temperature nears the set point the switch-off thrust is coming more from the sensing head in contact with the pan and less from the proportioning tube as the periods of current flow shorten. As the saucepan comes into the control temperature the control is mainly from the sensing head with just sufficient proportioning to prevent an overshoot. This proportioning is also sufficient to give a very close differential ($1°C$ or $1·5°C$ when good thermal conduction exists).

15

REFRIGERATION CONTROLS

THE refrigeration cycle is based on the fact that when a liquid evaporates it absorbs heat from its surroundings. This is due to the exchange of latent heat on evaporation, whereby the sensible heat of the surfaces surrounding the evaporating liquid is reduced. In a refrigeration circuit the vapour is trapped so that, after the process of evaporation and heat absorption, it can be retrieved, compressed, cooled, and used again. Two practical means of achieving this cycle are found in the absorption and compressor cycles.

The absorption cycle is essentially a totally-enclosed chemical process relying only on an external source of heat to set it in motion, this being controlled by an on-off thermostat.

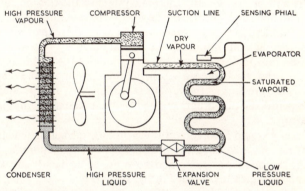

Fig. 15.1. Essential components in the compressor refrigeration cycle.

Compressor refrigeration

The basic compressor refrigeration cycle is shown in Fig. 15.1. Vapour is compressed in the compressor and is conveyed to the condenser where it is converted to a liquid as it cools. From the condenser the liquid passes along the pipe line to an expansion valve. This is a throttling device which regulates the flow of liquid into the evaporator, where it vaporizes as it expands and cools. The vaporizing process absorbs heat from the wall of the evaporator and, in so doing, extracts heat from the outer walls of the evaporator which, in turn, chills the cabinet in which it is sited. The expansion valve can take several forms including: (a) fixed orifice; (b) a float-operated liquid-level valve; (c) a straight pressure-operated valve; or (d) a thermostatically-operated expansion valve.

Thermostatic expansion valve

It is the thermostatic expansion valve with which we are concerned as a thermal device. Whilst the first three devices mentioned are suitable for small cabinets where the load factor is reasonably constant, for any application where the load factor is likely to change drastically the thermostatic expansion valve is the best solution. This device regulates the flow of liquid refrigerant into the evaporator, where the lower pressure causes it to evaporate and completely vaporize before it reaches the outlet of the evaporator. A temperature-sensitive phial connected to the expansion valve is sited on the suction line and is so arranged that on a rise in temperature in the line, indicating that insufficient liquid is being allowed into the evaporator coil, there will be a similar rise in pressure in the thermal system of the thermostatic valve causing it to open and increase the flow of refrigerant into the evaporator.

In its simplest form, the thermal system of the expansion valve is charged with the same liquid as that used in the refrigeration cycle itself. Hence, see Fig. 15.2, the pressures either side of the diaphragm in the expansion valve are balanced when the pressure above the diaphragm created by the pressure of the vapour in the sensitive phial is equal to the pressure below the diaphragm created by the pressure of vapour in the evaporator.

This theory would hold good if it was not for the spring which can be seen biasing and increasing the pressure below the diaphragm. This is known as the *superheat* spring and governs the degree of superheating of the vapour which takes place in the suction line.

One of the essential features of the thermostatic expansion valve is that it should prevent unevaporated liquid from passing through the evaporator and entering the compressor where it could cause mechanical damage. The thermal system in the expansion valve ensures this to some extent, because when the temperature of the suction line (i.e. the suction gas) drops, indicating that neat refrigerant is present, it shuts off the supply of refrigerant to the evaporator.

As an additional precaution the vapour is allowed to rise to a temperature approximately 5°C above its fully-saturated state and so to become superheated. This is achieved through the medium of the superheat spring, introduced beneath the diaphragm or bellows in the expansion valve. The thermal system now has to overcome the evaporator pressure at the inlet plus a measure of spring pressure, which will be equivalent to the additional pressure at the outlet of the evaporator caused by allowing the vapour to become superheated. Once this balance has been struck, the valve continues to function as a modulating device. A rise in temperature at the sensing phial will indicate that all the vapour has been boiled off and superheated. Further rise in temperature at the phial will cause a further pressure increase in the thermal

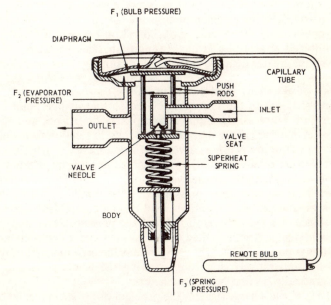

Fig. 15.2. Typical thermostatic expansion valve.

system, which will force the diaphragm or bellows downwards against the combined pressures of the evaporator inlet and superheat spring, and will allow further supplies of refrigerant to flow into the evaporator. On the other hand, if the sensitive phial senses a drop in temperature, indicating that the vapour leaving the evaporator has not been fully superheated, then it will produce a fall in pressure in the thermal system which will allow the diaphragm to contract taking with it the valve to the closed position. In this way, the valve can function continuously, regulating the supply of refrigerant to the evaporator in relation to changing load conditions around it.

Types of valve

In addition to the simple device just described, there are other more complex forms of thermostatic expansion valve for specialized applications. It is sufficient at this stage to refer to the various types of thermal charge that can be utilized in these types of valve; these include:

(1) *Liquid vapour-pressure charge.* These valves, as described in the last example, operate on vapour-pressure curves. Their design must take into account the problem of ensuring that the volume of liquid in the system is such that a surface of separation between gas and liquid is maintained at all times in the sensing bulb. This problem has been discussed in greater detail in the chapter dealing with prime movers.

(2) *Gas charge.* This is of the *limit-charge* or *fade-out* variety. The amount of liquid in the system is limited so that at one point in the temperature range all the liquid is completely boiled off, and further rises in temperature cause the valve to operate in relation to gas curves rather than saturated-vapour curves, i.e. further increases in temperature will only cause relatively small or negligible increases in pressure. This type of valve is used where the evaporator can attain high temperature and where a liquid vapour-pressure valve responding to this would open wide and put an undesirable load on the compressor during the early stages of a cycle. By ensuring that the pressure developed in the thermal system is strictly limited to the gas laws above certain temperatures, this is prevented, and the refrigerant is metered in to the evaporator at a comparatively slow rate, allowing the whole system to cool down gradually on first being switched on.

As with all gas-charge systems the problem of reversal must be taken into consideration. Should the bellows at any time become the coldest spot in the system, the comparatively small amount of liquid will condense at the bellows and transfer the temperature-sensitive point of the thermal system to this end. Therefore, a gas-charged thermostatic expansion valve must be fitted on a part of the system where it can be maintained at a higher temperature than the sensing phial. The same precaution has to be taken in respect of any capillary run, for if this at any stage was allowed to come in contact with the evaporator, then the valve would start working on evaporator temperature instead of the suction-line temperature.

(3) *Pressure-limiting valves.* In addition to limit-charging as a means of protection against abnormal working pressures, another type of valve is available which provides this safeguard by mechanical means.

Such a valve is shown in Fig. 15.3. In addition to the usual diaphragm which, under the influence of the charge pressure, regulates the flow of refrigerant to the evaporator, there is a second diaphragm one surface of which is exposed to the charge pressure and the other surface to an opposing spring pressure.

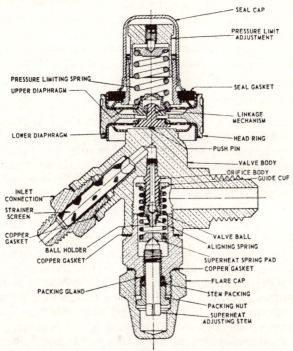

Fig. 15.3. Thermostatic expansion valve with adjustable pressure limit and adjustable superheat. (Controls Co. of America.)

Under normal conditions the valve functions as an ordinary thermostatic expansion valve. When under a high heat load a predetermined maximum operating pressure is reached, the valve's pressure-limiting feature, consisting of the second diaphragm and the limiting spring, takes over control of the refrigerant flow from the power element and throttles the flow of refrigerant to the evaporator to prevent suction pressures from increasing beyond the predetermined maximum level. This state continues until the overload condition has passed and pressures are reduced, after which the valve will again function as a thermostatic expansion device.

(4) *Adsorption charge.* This system has been described in greater detail in the chapter dealing with prime movers; basically it consists of a solid adsorbant substance in the sensing phial, such as activated carbon, and an inert gas. At low temperatures the gas is adsorbed into the solid adsorbant. Rises in temperature cause the adsorbant to discharge some of this gas and produce a pressure rise in the system. The thermal sensitivity is somewhat less than that of either the liquid or the gas-charge systems and this, in some cases, can be an advantage. Similarly, a wider range of operating temperatures is sometimes a desirable feature of expansion valves and the adsorber charge is capable of providing this.

Thermostatic switches

Whilst the refrigeration process is in operation, it is necessary for the compressor to work continuously, withdrawing superheated vapour from the evaporator, compressing it, and passing it to the condenser. This process continues until cabinet temperatures have reached the desired point, and then the compressor must be switched off. To accomplish this, a thermostatic switch is introduced into the circuit, and the refrigeration cycle is interrupted until the heat gain in the zone round the evaporator causes the thermostatic switch to make circuit and switch on the compressor, so restoring the refrigeration process.

Thermostats used in refrigeration work generally conform to the principles already described for on-off electric thermostatic devices, but have in themselves one or two requirements peculiar to refrigeration. The operating temperature range of a refrigeration thermostat is generally below ambient, a typical range being −12°C to 7°C, yet in normal storage, or when the refrigerator is not working, the same thermostat and sensing element can be subjected to high ambients up to 50°C. This means that any thermal system designed for a refrigeration application must incorporate features which enable it to be subjected to these high over-run temperatures without permanent damage. Similarly, whilst in normal operation the sensing bulb will be the coldest part of the thermal system, some applications create conditions where at some part of the cycle the thermostat head housing the diaphragm or bellows part of the thermal system could, in fact, be at a lower temperature than the sensing phial. This implies that the thermal system must be so designed that even if these reversal conditions occur, temperature control by the sensing bulb is still retained.

Two types of thermal system are in general use on refrigeration thermostats: vapour-pressure/liquid charge, and limit/gas charge. The vapour-pressure or liquid charge normally incorporates a sensing bulb on the end of a length of capillary which, in turn, is connected to a bellows or diaphragm in the thermostatic switch. The sensing bulb is partially filled with a liquid that produces a useful working vapour pressure over the range of the thermostat. The basic principles of these systems have already been covered in Chapters 3–5 and need not be repeated here. The problem of reversal is especially important in the application of vapour-pressure systems to refrigeration controls for, in many instances, the controlling head housing the flexible capsule can be the coldest point in the thermal system.

Limit-charge/gas-charge thermal systems

It has already been described how a vapour-pressure system can be charged with a limited amount of liquid so that at a certain point all the liquid is boiled off into vapour and where further increases in temperature cause this vapour to be superheated. The resultant increases in pressure follow the gas laws and, therefore, produce smaller increases in pressure per degree rise in temperature than is the case whilst saturated vapour remains in the system. In refrigeration thermostats this feature is frequently used and, in fact, the thermal range of the thermostat is entirely in the gas curve. In practice, the system is charged with vapour at a known pressure and temperature condition, this being controlled for the thermal range expected of the particular thermostat to which the sensing element is to be applied.

Apart from the protection they provide against high ambient conditions these gas-charged systems offer great sensitivity due to the fact that both the charge and the sensing phial containing it have a very low thermal mass. The only disadvantage of this type of system is that the gas will tend to condense at the coldest point in the system and, therefore, reversal conditions will apply if at any time the bellows or diaphragm in the thermal unit becomes the coldest spot.

Domestic refrigerator switches

The domestic refrigerator switch has undergone a drastic change during the last decade. In its early form it was a somewhat bulky assembly normally housed in the refrigerated zone. This called for very robust water-tight construction to ensure that the moisture in the refrigerated zone was prevented from entering the switch compartment and causing premature failure. Present-day designs for the domestic refrigerator overcome this difficulty by housing the switch outside the refrigerated zone, and having a length of capillary which is brought into the cabinet terminating at a phial which senses evaporator suction line temperature. The domestic switch invariably uses the limit or gas-charged thermal system, which offers the advantage that the sensing element need only be part of the capillary tubing since there is no liquid in the system in the normal temperature range. It is the practice to house the thermostat out of the refrigeration zone, and this has led to a demand for very small switches capable of being housed within the lining of the refrigerator, a typical design being shown in Fig. 15.4.

The domestic refrigerator thermostat is probably one of the most hard-working thermal devices in existence, and the materials used for

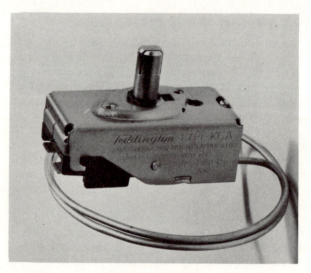

Fig. 15.4. Modern domestic refrigerator switch. (Teddington Autocontrols Ltd.)

its switch contacts need special consideration. The contacts are expected to make and break circuit many times per hour, day in day out, for many years without failure. To achieve this, the contacts are generously sized and manufactured in silver, with other metals, such as rhodium, added to give increased life. Great care is exercised in the design of these switches to ensure that contact bounce and chatter are eliminated even though the thermostat is called to operate under conditions of slight vibration from the compressor.

The operating differential of refrigeration thermostats varies widely with the type of application. This can extend from 1·7°C in thermostats called upon to control absorption-type refrigerators, up to differentials as wide as 5·5°C–11°C for compressor-type refrigerators.

Commercial and industrial types

Commercial and industrial refrigeration thermostats as used on cold-storage plant, etc., follow the same general pattern as the domestic switch, with the exception that limitations on size are not so important so that the thermostatic switch can be housed in a more robust case. The thermal system can take the form of the liquid-charge/vapour-pressure type of system or can equally use the limit/gas-charged system for increased sensitivity and protection against high ambients. The switching is carried out through silver contacts, the snap-action being derived either from a magnetic assembly or a spring operating through a dead-centre device. Where operation in conditions of high humidity or moisture is required, the mercury switch is employed as shown in Fig. 7.8. Here the thermal system rocks a bar on to which the mercury tube is mounted. The tube can take the simplest form of two electrodes shorted by the mercury, or more complex forms capable of providing changeover operation. Some manufacturers supply two mercury tubes in parallel so that two separate circuits can be operated at the same time.

16

COMMERCIAL AND INDUSTRIAL APPLICATIONS

INDUSTRIAL process control becomes increasingly complex as new processes demanding accurate control over many variables such as pressure, humidity or rate of flow in addition to temperature control are developed. A book such as this, dealing purely with temperature control, can only therefore go into this subject to a limited extent. Nevertheless in industry today there are many individual applications for temperature control which can be solved simply by the use of an on-off thermostatic switch or a direct-acting thermostatic valve. Such applications are discussed in this chapter.

On-off or two-position control

The discussion in earlier chapters of the advantages and limitations of on-off control has shown that this type of control offers a satisfactory solution to industrial temperature-control problems where: (1) some differential is necessary or can be tolerated; and (2) the total thermal lag in the system, comprising the effects of lag in the thermostat, the heating medium, and the controlled medium, is sufficiently low to react with the thermostat and still provide the degree of control required.

This method of control generally centres round an on-off electric thermostat which in turn controls heating or cooling media directly through an electric current, or moves a valve which, in turn, can control a flow of hot water, steam, gas, etc. Such a simple arrangement often proves adequate for the control of drying ovens, washing processes, and some curing processes where rapid heating up from a cold start is an essential feature and is more important than very accurate control once the correct temperature has been reached.

In a more refined form, on-off or two-position control can be adapted so that it offers some degree of graduated control of a process over a certain temperature band. This especially applies where the thermostat controls a valve which, in turn, controls the flow of gas, steam, or hot water. In the normal manner, the thermostat would switch either on or off, and move the valve from the fully-open to the fully-closed position, through the medium of a motorized damper unit or reversing motor.

But, by adjusting the linkage between the motor and the valve, it is possible to arrange that instead of moving the valve through its full stroke the motor merely moves it through part of its available movement.

For instance, where load conditions follow a certain pattern, calling for heating somewhere between a valve position of quarter open and three-quarter open, it can be arranged that the thermostat is able to move it only between these two positions. Additional temperature-limit switches can be added to the circuit so that if for any abnormal reason full heat is required, then the stroke of the motor-to-valve linkage is altered and allows the on-off control to revert to its full stroke enabling the valve to be fully-open while the abnormal demand for heat persists. Most industrial equipment is installed to perform a specific duty and in normal day-to-day operations the load factor remains reasonably constant. Hence this type of control can often provide adequate and inexpensive monitoring of the operation without the complexity of the proportional-plus-reset type of control.

Direct-acting proportional control

For a variety of processes using gas, hot water, or steam as a heating medium, direct-acting proportional control can be effected by a thermostatic valve, the thermal element of which contains a flexible member such as a diaphragm or bellows which exercises a direct throttling control on the heating medium. Such valves are frequently used in the control of water or steam-jacketed vats and heat exchangers.

With gas firing a typical example of this type of control is the rod-and-tube thermostat (see Chapters 1 and 13) controlling small gas-heated vats, or platens, the gas flow being modulated to maintain the correct heat input for a given steady temperature of the bath or platen. Such arrangements are frequently found in the control of die-vat temperatures and solder baths where the gas rate is comparatively low. The standard rod-and-tube thermostat is normally able to control flows of the order of 1 m^3/h of towns gas. For gas rates higher than this the thermostat is made to work in conjunction with a gas relay valve; this type of control provides a gradual shut down from the full-on to the full-off but it resembles on-off control rather than proportional action.

For the control of hot water and steam there are several direct-acting valves suitable, using either vapour-pressure systems or hydraulic-expansion systems. Fig. 16.2 shows a vapour-pressure control applied to a storage calorifier. The bulb of the controller is completely immersed in the calorifier and positioned away from the make-up entry. If an indicating thermometer is fitted to the calorifier, the bulb of this instru-

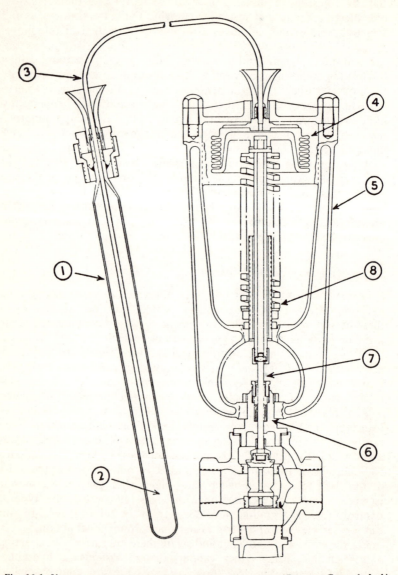

Fig. 16.1. Vapour-pressure-operated direct-acting thermostat. (Drayton Controls Ltd.)

1 Thermosensitive bulb. 2 Volatile liquid. 3 Capillary. 4 'Hydroflex' metal bellows. 5 Yoke. 6 Valve stuffing box. 7 Connecting spindle. 8 Temperature setting spring.

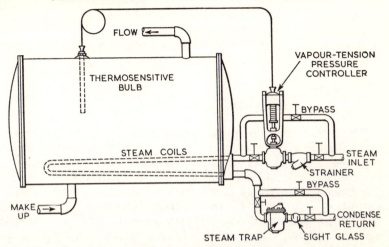

Fig. 16.2. Thermostatic control of a storage calorifier using a vapour-pressure operated direct-acting thermostat.

ment should be mounted in an adjacent position. The by-passing arrangement shown will permit the strainer to be cleaned, and the valve to be serviced without shutting down.

The principles and methods of operation of the vapour-pressure and hydraulic systems have been discussed at some length in Chapter 3. Each has its particular advantages and limitations for industrial applications. The vapour-pressure system is only temperature-sensitive at the bulb, and therefore is not affected by head temperatures surrounding the valve itself. On the other hand it only develops power in relation to the vapour-pressure in the bellows; hence when it is used to control single beat valves on high-pressure systems there is an upper limit above which the thermal movement is no longer able to oppose the unbalanced pressure within the valve itself. The hydraulic system can operate at much higher unbalanced pressure than the vapour-pressure system, but suffers from the possible disadvantage that the whole of the thermal system is temperature-sensitive; therefore, unless compensation is provided, some drift in thermal performance can be expected if the temperature of the bellows in the valve mechanism is subjected to some fluctuation.

Systems using direct-acting proportional control operate satisfactorily so long as the load factor remains reasonably constant. The valve assumes a certain position in relation to its seat at a predetermined temperature; therefore, so long as the flow through the valve in this position is such that it maintains the correct temperature condition in the heated medium, effective control will be maintained. If for any reason the load factor alters so that much more heat is required to maintain the same temperature then a permanent offset will result. The reasons for this are explained in Chapter 6.

Remote or servo-assisted control offering proportional, reset, or integral control or combinations of these

When the limitations of the control systems previously discussed constitute too severe a disadvantage, or when temperature control is only one of many factors to be fed into a more complex system, then thermostatic control becomes part of an electrical, electronic or pneumatic system feeding its information into a control circuit which exercises correcting action in relation to all these factors. In this type of control the thermal element can take the form of a thermocouple, electrical element, vapour-pressure system, hydraulic-expansion or bi-metal.

The control circuit will follow the principle of the Wheatstone bridge which has already been described in Chapter 8.

BIBLIOGRAPHY

BROOK, D. V. and BURKE, S. A., *Small Pipe Forced Circulation Central Heating*, British Coal Utilisation Research Association.

BURDETT, G. A. T., *Automatic Control Handbook*, Butterworths (1962).

COXON, W. F., *Temperature Measurement and Control*, Heywood and Co., Ltd.

EGGINTON, H. E., *Refrigeration Controls*, Refrigeration Press Ltd.

HAINES, J. E., *Automatic Control of Heating and Air Conditioning*, McGraw Hill Book Co. (1961).

JONES, E. B., *Instrument Technology*, Vols. 1 to 3, Butterworths (1965, 1956 and 1957).

LeFEVRE, R. N., *Domestic Utilisation of Gas*, Walter King, Ltd.

WIGHTMAN, E. J., *Instrumentation in Process Control*, Butterworths (1972).

Code for Temperature Measurement, BS 1041 (1943).

Glossary of Terms Used in Automatic Controlling and Regulating Series, BS 1523: Part 1 (1967) and Section 5 (1954).

Instrument Manual, United Trade Press, Ltd.

Room Thermostats, BS 3955: Section 2F (1967).

Special Study of Domestic Heating in the United Kingdom, Institute of Fuel (1956).

Thermostats for Gas Burning Appliances, BS 4201 (1969).

Thermostats for Use with Domestic Electric Cooker Ovens, BS 3955: Section 2E (1966).

MILES, V. C., *Domestic Vaporiser Burner Practice*, Oil Firing, London (1961).

MILES, V. C. and PINKESS, L. H., *Gas Appliance Controls and Practice*, Earnest Benn, Ltd. (1970).

INDEX